Magic Science Religion

Experimental Practices

TECHNOSCIENCE, LITERATURE, ART, PHILOSOPHY

General Editors

Sher Doruff
(*Gerrit Rietveld Academy*, and *Amsterdam University for the Arts*)
Manuela Rossini
(*University of Basel*, President and Executive Director of SLSAeu)

Editorial Board

Yves Abrioux (*Université de Paris 8*)
Jeoren Boomgaard (*Gerrit Rietveld Academy, University of Amsterdam*)
Elisabeth Friis (*Lund University*)
Stefan Herbrechter (*Coventry University*)
Vicki Kirby (*University of New South Wales*)
Laura Otis (*Emory University*)
Lynn Turner, Goldsmiths (*University of London*)

VOLUME 2

The titles published in this series are listed at *brill.com/exp*

Magic Science Religion

By

Ira Livingston

BRILL
RODOPI

LEIDEN | BOSTON

The Library of Congress Cataloging-in-Publication Data is available online at http://catalog.loc.gov
LC record available at http://lccn.loc.gov/2017049984

Typeface for the Latin, Greek, and Cyrillic scripts: "Brill". See and download: brill.com/brill-typeface.

ISSN 1873-8788
ISBN 978-90-04-35710-5 (paperback)
ISBN 978-90-04-35807-2 (e-book)

Copyright 2018 by Koninklijke Brill NV, Leiden, The Netherlands.
Koninklijke Brill NV incorporates the imprints Brill, Brill Hes & De Graaf, Brill Nijhoff, Brill Rodopi, Brill Sense and Hotei Publishing.
All rights reserved. No part of this publication may be reproduced, translated, stored in a retrieval system, or transmitted in any form or by any means, electronic, mechanical, photocopying, recording or otherwise, without prior written permission from the publisher.
Authorization to photocopy items for internal or personal use is granted by Koninklijke Brill NV provided that the appropriate fees are paid directly to The Copyright Clearance Center, 222 Rosewood Drive, Suite 910, Danvers, MA 01923, USA. Fees are subject to change.

This book is printed on acid-free paper and produced in a sustainable manner.

This book comes from my commitments to the study of complex systems, to philosophy and cultural theory, to poetics—and from a refusal to choose among them. Those, in any case, are some of the hands on the planchette. *In the pages that follow, you'll find the message that they've spelled out.*

My role model in this might well be Erasmus Darwin—physician, philosopher, abolitionist, and poet—whose scientific epic poem The Botanic Garden *(1791) and strangely poetic treatise* Zoonomia *(1794) prefigured evolutionary theory well before his grandson Charles came along and put it on a more solid footing.*

So, if by serving my multiple commitments I have failed to do justice to each, and failed more necessarily and fully to materialize the three-headed chimera named by the title, I can say, at least, that I don't think it's a quixotic undertaking. I hope my grandchildren can do better. This book is dedicated to them and that.

∴

Contents

About this Book XI

1 **Introduction** 1
   1   All or Nothing? 1
   2   Kindness 2
          ASIDE: The Two-Million-Dollar Comma 2
          ASIDE: Sufficiently Magical Technologies 3
          ASIDE: Our Gods 4
   3   Cruelty 6
          ASIDE: Ghostly Causality 7
          ASIDE: Durable Cruelties 8
   4   Beginning, Again 14
          ASIDE: Chaos Invoked 14
   5   The Freedom Not to Choose 15
          ASIDE: No Atheists in Foxholes 16
          ASIDE: Blake on Firm Perswasion 17
          ASIDE: Metafictional Magic: The Writer's Will 19
          ASIDE: Shapely Sentences 21
          ASIDE: Beckett Meets *Twilight Zone*, Glass-Half-Full Version 23

2 **Complex Systems in a Nutshell** 25
   1   Horror Movie Reboot 25
   2   Interpositivity 25
          ASIDE: Interpositivity 27
   3   Becoming a System 27
   4   Creatures of Light 29
          ASIDE: Transformative for Whom? 30
          ASIDE: Plant Sorcery 31
   5   Tornados, Whirlpools, and Fires 32
          ASIDE: iii 33
          ASIDE: Co-evolution 34
   6   Leveled 35
          ASIDE: On being born too early and dying too late 36
          ASIDE: Up Around the Bend 37
   7   A Personifying Universe of Stretchy Things 38
   8   Dynamism 40
          ASIDE: One Plus One Equals Infinity 41

9   Magic, but No Black Boxes   41
          ASIDE: The T-Shirt Version   43
  10   Wildness   43
          ASIDE: Pet Resemblance via Social Theory   44

3 **Magic by Example**   46
  1   Failed Magic: Modernist Heaven (and Hell)   46
  2   Placebo Magic   50
          ASIDE: Pregnancy via Magic   52
          ASIDE: Following the Scent   53
  3   Mirror Magic   54
  4   Biting Game   59
  5   Dog Whisperer   61
  6   Conclusion   62

4 **Future Perfect**   63
  1   The Gettysburg Address as a Magical Speech Act   63
  2   Pool, Poetry, Prose, and Painting   64
          ASIDE: Dr. Livingston's Magical Bank-Shot Visualizer   65
  3   Meteors, Messiahs, and Migraines   69
          ASIDE: On This Rock I Build My Church   69
          ASIDE: Falling Off a Log   71
  4   Magical Militarism   72
  5   Four Asides   74
          ASIDE: Dr. Livingston's Time Travel 101   74
          ASIDE: A Somewhat Rationalizing Account of Mystical Nonviolence   76
          ASIDE: A Somewhat Mysticizing Account of Rationalist Nonviolence   77
          ASIDE: Four Scholars   78

5 **What is Religion?**   82
  1   A Fuzzy Set   82
  2   Magic, Science, and Religion Coevolve   83
  3   Restoring the Chaos   84
  4   Rain Dance   87
          ASIDE: From Flow   89
  5   Providence, via Vico   91
          ASIDE: Systematizers Systematized   93

6 **God 3.5B (A Nearsighted Evolutionary Panorama)**   97
  1   Salience to Sentience   97
  2   Primordial Systemhood Membership Narratives   100
          ASIDE: Law and Order   101

	3	The Myth of the Acquisition-of-Consciousness Moment 103
	4	Use of the Terms *I* and *We* 104
	5	Cis-Systems → Trans-Systems 105
	6	First Caveat: Characters 106
	7	Plant and Animal Intelligence 107
	8	Learning and Evolution from a Systems Perspective 108
	9	Second Caveat: Plot 111
	10	Especially Informative Postinfundibular Amphioxus Hypothalamus Neuropile 114
	11	The Holy Grail 120
		ASIDE: Neuro-X 121
	12	Teamlikeness 124
		ASIDE: Play and Teamlikeness 125
		ASIDE: Three-Ring Circuits 126
	13	Conclusion 136

7 **Blake Magic**   138
   1. Famous in Heaven   138
      ASIDE: No Success Like Failure   139
   2. Blake Magic   140
   3. Auguries   143
      ASIDE: Prophecies   145
   4. Angel Kings   146
   5. Magic in the Current Conjuncture   149
   6. Five Asides   154
      ASIDE: An Elegant Pose   157
      ASIDE: Ritual Retraction   158
      ASIDE: Obscurity   160
      ASIDE: Butterfly Tornado   161

Acknowledgments   165
References   166
Index   173

About this Book

How big is a thought? How many elements make one?
Where does it go when it gets too big for a sentence?

As I've said, I set out to explore the unlikely intersections
of magic, science, and religion. Somewhere along the way,
I started organizing the text into three- and four-line units.

I wasn't out to create a hybrid form. It's simpler than that.
I've never understood paragraphs, never thought in them.
I tried to suppress the smaller units, reorganizing the text
into paragraphs, but the short units reasserted themselves.

I claim no integrity, necessity or exemplariness for them.
I do notice that, in describing them—something like
crosses between sentences, paragraphs and stanzas—
I replicate the notion of the "unlikely triple intersection"
named in the title—and that, like the systems I describe,
the book's whole and parts emerged and evolved in tandem.

If it bothers you, just think of it as prose. *It's just prose.*

So, quickly to situate this quirky, formalist experimentalism
back into a world where it would seem to be a reckless luxury:
here we all are, end times and early days. Flashback:
sixty-six million years ago. Some of the mutants may survive—
after all, the more the merrier, the smaller and weirder the better.
Though history will little note what we write and read here,
some of the DNA may survive. *Can anyone say which?*

Here I am, on the edge of prose, flapping my not-yet-wings.

CHAPTER 1

Introduction

1   **All or Nothing?**

Magic, science and religion, together, add up to more
than can be covered in a lifetime, much less a single book.

What interests me, though, is not the sum but how they intersect,
and since the three are usually defined in mutual opposition,
we go from something that's too huge and sprawling to map
to what might be a *null set*, an intersection with nothing in it.

To start small, can we start by looking at promising examples—
candidates, at least, for membership in this intersectional set?
Might this transform how we engage magic, science, and religion?

I don't want to seem to promise something I can't deliver—
so, almost by definition, none of the examples can be *pure*:
neither the most magical, most religious, nor most scientific.

If you want to find the intersections among things
that you've been trained to put in different categories,
look at the margins. In so doing, you gain something.

Intersectionality has powers that centrality and purity lack.
Here on the shakiest ground, where tectonic plates meet,
you come to understand the system as a dynamic one.
You find *the leading edge of its aliveness*, of its evolution.

For a start, the marginal intersection starts to get repopulated
with a quirky menagerie. And we're not outside, looking in.
This is where we live.

We have to do some groundwork to recognize such examples,
but I have liberally mixed the groundwork with examples, too.

The journey is necessarily a crooked one, with detours and asides.
The winding path—the path on which we may repeatedly get lost,
traverse switchbacks, wander and wonder if we're going in circles—
is a way to get out of familiar, uninspiring territory we know too well.

2   Kindness

I am interested in how small actions—small to the point of insignificance—
can have big effects, even at several circuitous removes from what they affect,
and even though they operate in a different dimension from what they affect.

If you figure out how to do this on purpose, it warrants being called *magic*.
I approach this from the study of systems in relation to their subsystems,
other systems, and environments, but I've kept it as nontechnical as I can.

Whatever more exotic actions fall under this description
(we'll be considering some examples in the coming chapters),
it also fits the everyday operation of *language and writing*.

The symbolic strings we call language (such as sequences of 26 letters,
augmented by a few other symbols, that here comprise written language)
would matter almost not at all to the world at large. The differences
among *Mein Kampf* and "On the Electrodynamics of Moving Bodies,"
Dr. Seuss and *The Holy Kabbalah* signify next to nothing in the world
except to humans—but *through us,* these rearranged sequences of letters
have dramatically different repercussions for the worlds we make.

If you get beyond the goes-without-saying obviousness of this,
you can find the terrain of real magic, hidden in plain sight.

> ASIDE: The Two-Million-Dollar Comma
>
> The 1870 U.S. Tarriff Act exempted from import duties "fruit plants, tropical and semi-tropical, for the purpose of propagation or cultivation."
>
> When the legislation was renewed, someone inserted another comma ("Fruit, plants, tropical and semi-tropical"), costing two million dollars in lost tarriffs (which, adjusted for inflation, translates to $38 million).

This may be one of the *least magical* examples of the power of language!
Even so, the comma must have had real economic and ecological impact
on the cultivation and movement of actual fruit and plants in actual space.

The comma works through a Rube Goldberg, smoke-and-mirror series
of consensual and highly artificial human fictions (virtual realities)
that include language, money, laws, and national boundaries.

Only via virtual worlds we've created and woven so finely
and so broadly across the planet could something so small
be magically empowered to have such large repercussions!

For 100,000 years, language has been wired and rewired
back through human brains so deeply and so intricately
that our collective world would collapse without it.
Through language, a thought (a subtle set of electro-chemical events
in the brain) can change the world. And this is a routine occurrence.

ASIDE: Sufficiently Magical Technologies

When French sailors met the Yamacraw Indians
(in what's now Georgia) in 1562, it was the first time
the Yamacraw had encountered written language.

They were wowed "that this white Man could send his Thoughts
to so great a Distance upon a white Leaf, which surpassing
their conception, they were ready to believe more than a man."

Science fiction writer Arthur C. Clarke's famous maxim that "any
sufficiently advanced technology is indistinguishable from magic"
might be applied here, even to technology so routine to us as writing.

The maxim is often regarded as a truism, in spite or *because* of
its smug rationalism, annoying aggrandizement of technology,
and condescension toward "primitives." To be fair, it suggests that,
facing a more advanced civilization, we'd stand in the same shoes.

But *our* problem, as rationalists, is not, as Clarke implies,
a failure to distinguish technology from magic. I don't think

of my computer and smartphone as magical deities, in spite
of having only the vaguest idea how they work, and in spite
of digital technologies relentlessly being marketed as magical.

Even so, regarding them as deities—as an anthropologist
studying me might conclude that I do—might well generate
the best account of how we've become enslaved to them!

> ASIDE: Our Gods
>
> In his 1788 book *Aphorisms on Man*,
> Swiss theologian J.K. Lavater proposed
> that what Christ "meant when he said,
> WHERE THY TREASURE IS,
> THERE WILL THY HEART BE ALSO,"
> is that *"the object of your love is your God."*
>
> Alongside these words in the margins of his copy
> of the book, William Blake responded: *"This
> should be written in gold letters on our temples."*

The much more important and productive task, then,
is *the reverse* of what Clarke suggests: to come to terms
with what *is* magical even about routine practices
such as writing. This coming-to-terms is vital
when you want to find or enhance the magic, and
also when you want to resist it. Hence this book.

One way to frame the difference between magic and science
is by the proposition that *magic is the art of producing
changes in consciousness through exercise of the will.*

The aim may be to influence the world via consciousness,
but *science* (we are told) *aims to change the world directly.*
However, *science is also wired through brains* (and bodies).

In fact, science is an exemplary case of the occultish powers
that can be attained when brains are systematically wired
together over many generations, as in the case of *language*.

If you believe that light would travel at the same speed
regardless of whether we had measured it or not,
and that it travels at that speed now as it did in ancient Greece,
your belief is still specific to your moment in history.

But I want to bypass arguments about how facts are *invented*
versus merely *discovered*, how much facts are shaped
in cultural and theoretical and political contexts, and so on.

So in any case, it is clear that *our knowledge of the speed of light*
and *technologies of making things with light* come from the world
only via the collective brain wiring we call science and technology.

Sociologist Bruno Latour calls this *Nature Two*: not
facts posited as independent of us (*Nature One*),
but as established by a social, interactive process.

(And by the way, the Yamacraw were right.
They knew what they were up against; they saw
the writing on the wall. Can we say the same?)

The same virtually occult power can also characterize feelings—
subjective phenomena so deep that even a life-defining feeling
might register objectively only in a fleeting micro-expression,
or so patient as to come to be embodied in facial dispositions
that only fully manifest themselves over the arc of a lifetime.

Emotions and feelings—often grouped under the term *affect*—
operate to weave together subjective and objective worlds.

Affect connects a particular present (an assessment of what's going on in the moment)
with a past (via memories or by subconscious programming laid down by experiences of
similar situations) and a future (through anticipations of possible consequences).

The futurity in affect is why psychoanalyst André Green
described emotion, at bottom, as the "anticipation of a meeting
between the subject's body and another's body, real or imaginary."

For better and worse, affect stubbornly holds worlds together across time,
a form of magic. Change the affect, and the shape of spacetime changes.

If you can figure out how to intervene—at an individual or collective level
or at a point that links the two—to alter the way pasts, presents, and futures
are linked by affect, you can change the world. Some who do this routinely
are psychoanalysts, politicians, artists, and mothers: it's called *affective labor*.
Thoughts, language, and affect shape our worlds by *meaning*.

Is meaning (1) something we manufacture in an otherwise meaningless universe,
(2) something coming from God or the occult, framing and infusing everything,
or (3) the ways things have co-evolved to connect with and matter to each other?

We will explore many examples, but just a hint: it's Number Three!
If you like another answer, don't worry: when you follow it through
rigorously enough, it turns out to be a lot more like *all of the above*.

Language, thoughts, and feelings bear *withness*—
family resemblances—not only with each other
but with all kinds of processes and systems in the world.

This kinship—the quality of being the same *kind* of thing
(which is, for a start, what makes it a form of *kindness*)—
connects all in a *general ecology* that links mind and world.

This is where magic, science, and religion meet—and when I say *this*,
I also mean *this very interaction* between these words and your brain,
my brain and yours and all the other brains language plugs us into.

3   Cruelty

In a 1972 talk, climatologist Edward Lorenz first described the *butterfly effect*.
Even the beginning of the term's rise to paradigm status was circuitous.

When Lorenz failed to provide conference organizers with a title,
one of them called his paper "Predictability: Does the Flap
of a Butterfly's Wings in Brazil Set Off a Tornado in Texas?"

The point of the thought experiment was that, even if you covered the globe
with a grid of weather stations every ten feet, you still couldn't predict the weather
very far into the future, since even tiny fluctuations can have repercussive effects.

Butterflies would still fly under the radar.

Such causality *almost* qualifies as magic. The trigger
is so small and the distance between cause and effect so large
that it might be magic *if only the butterfly knew what it was doing*.

If fluttering butterflies blithely trigger tornados, where does it end?
Wouldn't the tornados have equally subtle but repercussive effects,
leading circuitously, in turn, to more repercussive series of events?

These might include the rise of fascist dictators, the wrong two people
coming together and giving birth to an Antichrist, finding a cancer cure
that unleashes a plague (but some of the mutants have superpowers).

All because of that cartoony, carefree Brazilian butterfly! Should we start
a global campaign to stamp out butterflies? Or, as someone in this film
is bound to ask: What if lepidopteracide *brings about* the catastrophe?

> ASIDE: Ghostly Causality
>
> Ray Bradbury's popular sci-fi short story "A Sound of Thunder"
> was published in 1952, twenty years before Lorenz's lecture.
>
> In the story, time travelers go back to prehistoric times. One of them
> accidentally steps on a butterfly. Upon returning to their own time,
> they discover that history has been disastrously altered.
>
> As far as I know, Lorenz never said the story contributed to his idea
> of the tornado-triggering butterfly, but it could have influenced him
> even if he'd read it or been told of it and then forgotten that he had.
>
> It could even have played some role *if he'd never read it*
> but just picked up on its widespread popular-cultural echoes.
> The butterfly-effect idea thus *performs itself* in its inception.
>
> We cannot trace the tornado back to the butterfly
> or the butterfly forward to the tornado amid the myriad
> other factors that had to align for the tornado to emerge.

> We will never know how much the science fiction writer may or
> may not have influenced the scientist, or by how winding a path.
>
> Maybe Bradbury and Lorenz were just elaborating the word *bug*,
> meaning *a glitch that triggers cascading failure in a machine*—
> a 19th-century usage that draws on much older sources.
>
> Middle English *bugge*, Welsh *bwg*, and Scottish *bogle*
> all mean *ghost* or *goblin*. The entanglement of local
> and system-wide causality is a key mode of *meaning*.
>
> Ideas and meanings—like butterflies and tornados,
> *and yes, ghosts*—are often emergent localizations
> of what may well already be "in the air."

You can imagine a form of mental illness in which you believed
that otherwise insignificant actions, even certain thoughts or feelings
could have disastrous consequences in the world at large.

This might lead you anxiously to police your thoughts and feelings,
repeatedly performing certain rituals in order to prevent this.
And there is such an illness: obsessive-compulsive disorder.

The premise of disproportionate influence isn't all wrong,
but systems emerge and evolve to be relatively stable,
even as they ride on seas of fluctuating indeterminacy.

The body can survive the "thousand natural shocks
that flesh is heir to" (as Macbeth put it), but even so,
a tiny air bubble injected into a vein can kill you.

It's clear that air bubbles don't often find their way into our veins.
If they did, creatures with veins would have evolved ways
to survive such an event—or never would have appeared at all.

> ASIDE: Durable Cruelties
>
> Human systems of meaning are durable in this way too,
> often persisting in spite of what you would have thought
> to be the most compelling evidence of their inadequacy.

INTRODUCTION

> Although religion and technoscience have, each in their own ways,
> kept people alive and helped them thrive, each has also killed them—
> and continue to conspire together to kill them—by the millions.

All of this points to something wrong with common understandings
of the "many worlds" interpretation of quantum theory:
if there's one universe where the butterfly causes the tornado
(or to say it carefully, contributes something crucial to the initial conditions
that enable the tornado to emerge), and a billion others where it doesn't,
the odds are still only a billion to one that it will happen.

This suggests that simply *recognizing* the thought or feeling that can rock the world—
among the clouds of them flapping around at any given time—may be the magic.

If Lorenz's discovery qualifies as world-rocking, consider that it occurred to him
when he reentered some data into a weather-simulating computer program,
rounding off, for convenience's sake, six decimal places to three.

When the second simulation began rapidly to diverge from the first,
he could have just sighed, reentered and rerun the original data—
and the butterfly effect would have fluttered by unnoticed!

In 1935, physicist Erwin Schrödinger
devised the famous thought experiment
on which the "many worlds" idea is based.

Imagine a box rigged so that a quantum fluctuation—
the tiniest event imaginable—will trigger a mechanism
to release hydrocyanide gas, killing a cat in the box.

Because of the indeterminacy that characterizes such fluctuations, a quantum particle
can effectively be in two places at once—until an observer intervenes to measure it.
Thus it seems the cat could be simultaneously *dead and alive* until someone checks.

In fact, living creatures have evolved to ride on top of such fluctuations:
we've selected which kinds of quantum events will signify and which will not.

For example, our sense of smell—important in an evolutionary sense because it helps us
recognize and bond with our loved ones and determine what food is safe to eat—
may sometimes depend on differences among quantum states of chemical compounds.

More broadly, all of biochemistry (the set of building blocks for life) is built
selectively on biophysics (the propensities of matter, as at the quantum level).

As complex systems, we live so much in worlds of our own making
that millions of other subatomic particles—ones we haven't selected
as significant—zip through our bodies constantly without us noticing.

Schrödinger's thought experiment goes against this grain.
While selected quantum events have *meaning* for creatures
in an ongoing, constructive, collective, evolutionary sense,
he arbitrarily binds a single life to a one-off quantum event.

To make a life depend on something it has otherwise evolved to ignore
requires a devious apparatus: Rube Goldberg meets Hannibal Lecter.
Someone who constructed such apparatuses would be *a sociopath*.

One could imagine a movie—a whole series of bad movies, in fact—
in which serial killers constructed elaborate apparatuses to kill people like this.
And you have heard that serial killers start with housepets such as cats, right?

In fact, not long after Schrödinger's proposal, his fellow Germans
began using hydrocyanide, the gas he chose for his thought experiment—
under the brand name *Zyklon B*—for mass extermination of Jews.

Another binding-back of humanity's fate to the behavior
of subatomic particles wasn't far behind: the atom bomb.

The cruelty is a *clue*. Why didn't Schrödinger imagine instead a scenario
in which complex order might have been created instead of destroyed,
such as a process that might have united an egg and sperm?

While Schrödinger thought of the dead/alive cat as an impossibility,
the experiment suggests to some how *observation* or *consciousness*—
"higher" phenomena—are wired back through "lower" substrates
(such as quantum particles and neurons and ink and cats in boxes)
in shaping ways you might call *magic* (though often lost on us),
as through *information,* that dead-and-alive, thing-and-non-thing,
baby-and-bathwater sequence of events by which my brain arranged—
in view of all these witnesses—the stuff that forms these words.

INTRODUCTION

Life emerges *from* as it is wired back *through* its substrates,
including quantum events (by selecting which will matter),
just as language emerges from and is wired back through
brains, bodies, social interactions, and the world at large.

In fact, the word *religion* originally meant a *binding back*,
apparently to particular sets of laws and commandments.

I broaden that concept to include the embrace of constraint as creative,
the way a result can become a cause, or how a higher level can become
part of the essence or origin of the lower level from which it emerged.

This broadened understanding of *binding back* is key
to the magic/science/religion intersection at the heart of this project.
To put it another way, I'm interested in magic that is *generative*.

When it comes to wiring the fate of humans back down to subatomic particles
in a possibly generative way, consider the Internet, with its flea circuses of photons
carrying via optical cables many Libraries-of-Congress' worth of data every day.

The jury's still out as to whether that ingenious bit of rewiring
may also be said to be making us simultaneously dead and alive.

In any case, in contrast with Schrödinger's scenario, the tornado posited
in the title of Lorenz's paper—an emergent subsystem of the weather,
shorter lived and more volatile than other weather patterns but also
somewhat more coherent—is recognizeable as *something like an entity*.

Like other living things, it exhibits a paradoxical mix of *overdetermination*
(requiring a host of factors to be precisely attuned for it to emerge at all)
and *underdetermination* (tabulating all factors may still not enable you to say
which way it will jog, whether it will break up or get in a groove and go to town).

Partly because the stakes are so high and little differences so stark,
tornados seem to have a *livingthinglike* or even *godlike* agency—
and we smaller gods are always at the mercy of their whims.

Could you ever learn to recognize the subtle pressure points of a weather system
so well you could perform magic: wave a wand and cause (or prevent) a tornado?

Lorenz's conclusion is that you couldn't—just as you couldn't predict
forward from butterfly fluctuations, or trace causality backwards to them.

It's the same reason why even the best pool player
can't reliably hit multiple-ball combination shots
when the number of balls involved gets too large.

The margin for error rapidly gets smaller than the *noise*
introduced by vibrations, irregularities of the table surface,
less-than-perfectly-spherical pool balls, and other factors.

Even a massive computer array—somehow supersensitively attuned
to everything in the immediate environment—couldn't predict exactly
how, say, a subway train passing underneath will vibrate the table.

Or even if it could, it still couldn't anticipate that—
after you've hit the cueball—some volatile person
might suddenly slam his fist on the next table over.

Turning to more generative or emergent processes,
think of the tiny permutations of DNA on which so much
of our lives depend. These we can already adjust at will.

But anybody can *screw things up*, and there are many words in Yiddish
for doing so. I'm not usually one to coin new terms, but for *Rube Goldberg
meets Hannibal Lecter*, we might try using the term *Schrödinger*.

Even with current knowledge, some Schrödinger could, by tinkering with DNA,
manage to cause assorted horrific mutations (perhaps already has done so)—
or well-meaning improvements with disastrous longterm repercussions.

Rather than simply intervening, can we ever master
the gothically complex contingencies and repercussive backlashes—
the feedback loops-upon-loops of DNA's many dancing partners?
Can we master them well enough to learn how to intervene in DNA
to enhance the ecology of humans in their complex relationships
with each other and the planet? That would be real magic.

Consider another famous thought experiment featuring an invented entity
known as *Maxwell's Demon,* who lurks at a gate between two spaces

equal in temperature and pressure (at thermodynamic equilibrium)
and closed in perfect isolation from the outside world.

When a fast particle comes toward him from one side, the demon
opens the gate and lets it though, but if a slow particle comes at him
from the same side, he closes the gate and lets the particle bounce back.

The demon reverses that protocol for the space on the other side. Soon, one side
begins to be dominated by fast particles—it heats up—and the other cools down.
While operating in a closed system, has he increased order and decreased entropy?

It seems the Second Law of Thermodynamics—which says entropy
can only *increase* over time in a closed system—has been violated!
Something seems to have been created from nothing.

But there's a problem. In acquiring the necessary information
about the particles, and in opening and closing the gate,
the demon would have had to draw energy from the system.

Because energy is orderly, becoming disordered as it gets used
(as when vented as heat), the demon couldn't increase the orderliness
of his system—unless energy was continually pumped in from outside.

That's exactly what happens when sunlight streams into an otherwise closed
aquarium or terrarium—and as it streams into the terraqueous bubble we call Earth.

If there isn't something for nothing, it seems that order can be increased
only at the expense of creating disorder around it. And if that's the case,
how do we go in the direction of an aliveness that isn't just local?

Serving one's own aliveness—say, by living (with a few dear friends, of course)
in a technobubble sustained by an army of slaves, harvesting organs from the poor—
is bound to create a catastrophic backlash—after, who knows, maybe a thousand years.

More immediately, what serves the aliveness of my thoughts and words, their ability
to insinuate or ingratiate themselves (like memes and viruses) with their human hosts,
to stay alive by being read, by continuing to dance with actual living neurons?

A longterm strategy for survival must involve serving the subsystem in question
and the ecosystem from which it emerges and with which it remains imbricated:

its "holding environment" (to borrow a term from psychology). This book is interested in the magic by which we emerge and evolve and stay alive.

4   Beginning, Again

ASIDE: Chaos Invoked

In the beginning, there is chaos. Restless play;
relentless, anarchic plenitude. The indeterminate
quantum hum that underlies everything.

Bubbling up, falling all over itself, getting in its own way,
going around and up and over itself, working itself into
frenzies and froths and exhaustions, flotsams and jetsams.

Smithereens cohere into pulsing planets. Continents
of solid rock flow and buckle, only slightly less volatile
than the seething life that comes to cover them in turn.

Overflows begun as trickles become torrents.
Glaciers push forward and recede. In their wake,
scattered campfires proliferate and ramify
into superclusters of electric light, and then,
just as quickly, wink out in waves of darkness.

(There, did you catch it? The history of humankind
in that last sentence?) In the beginning, there is chaos.
And there still is. Chaos, and beginning. Even here.
These words (I would like to think) are their agents.

To begin, then, again: after this introduction, we will explore magic, as it intersects with science and religion, via several main examples and a host of smaller ones.

The main examples will include a medical study of placebos, a treatment for phantom limb syndrome, a therapist curing a sick infant, a dog whisperer, the Gettysburg Address, and a two-hundred-year-old poem.

Because I study and practice writing, I focus disproportionately on magic as enacted through language. Even if there could be universal magic, there are no universal magicians.

If you want to do magic, you have to practice something
in an intense and sustained way—whether physics, cabinetry,
or even loving one particular person for a long time.

Being a magician in one realm doesn't make you competent in another,
but you can find subterranean or intramural passageways, wormholes
between and among the realms. That's what we're engaged in doing here.

5   The Freedom Not to Choose

To begin with a cliche: the world in front of you—
the world in which you are reading this sentence—
could be living, breathing, pulsating with magic.

It could be glowing—or circuit-blowingly overflowing—with *meaning*,
or revealed as empty of it. As many of us know from experience,
each of these states can be terrifying, depressing, or exhilarating.

Magic and meaning and affect are intimately related.
It may be that you can't have one without the others.

The world in which we find ourselves might be (1) churned out constantly
by impersonal laws and algorithms, (2) ruled by a stern or benevolent God,
or (3) sustained by the play of subtle, living constellations of entities.

If there is some choice (okay, it's Number Three again) as to which worlds
we recognize ourselves as inhabiting, and if there are life-changing
differences among them, how could anyone be blasé about it?

But *choice* is generally a consumerist ideology, pushed hardest by those
who want to *control* our choices. We're offered petty freedoms—wiggle room
to distract us from the ever more thoroughgoing constraints in which we live.

So we are going to attempt here an exemplary magic act: to be confined
by multiple sets of constraints—handcuffed and chained and locked in a box,
sunk in a tank of water—and somehow wriggle out, in view of witnesses.

We can't return to an enchanted world of magic, animated by spirits.
It's said that modernity destroyed such a world, or, more likely,
invented it retroactively, as a form of nostalgia for what never was.

But there's another reason why we can't return to such a world: We *never left*. We've been told that modernity sold us a clockwork universe, but the clock sits on the shelf next to the Buddha and the Virgin Mary and other deities, fetishes and oracles. We consult each as the occasion requires.

> ASIDE: No Atheists in Foxholes
>
> Trapped in a hole half a mile deep, 33 Chilean miners
> survived for more than two months after a mine collapse.
>
> They were collectively buoyed by a religious ardor
> that began with a simple numerological coincidence.
>
> It occurred immediately to the 33 men who made it
> to the emergency refuge when the mine began to collapse
> that Christ was 33 when crucified, buried, and ressurected.
>
> All were galvanized by daily prayer sessions, after which
> they would receive their tiny portion of emergency rations.
>
> The cache contained enough to feed 25 men for two days,
> but they survived more than two weeks with no casualties—
> something like a real loaves-and-fishes kind of miracle.
>
> "God exists!" one of them proclaimed when a drill finally broke through
> from above on Day 17 (though they'd spend 52 more days underground).
>
> Soon food began to be sent down to them. It became clear
> they'd be rescued—and in fact, that they were already celebrities.
>
> Predictably, their religious ardor began to cool. There's no denying,
> though, that it had served their individual and collective survival.

We tend to live not in one or another of these worlds but at the place where they overlap. This ensemble of worlds is shifty but not all muddled together, plural but not infinite. It is characterized by *someness*: a shifty and innumerable-but-finite plurality.

An exemplary case of someness is the human brain: "a complex system made up of many heterogeneous, overlapping, interacting and densely connected subsystems."

INTRODUCTION

So by wriggling (a way of describing the thought process of this book),
the aim is not to *move* so much as to *remain* at the magic/science/religion
intersection; to avoid being pulled into one at the expense of the others.

Especially when we are sold freedom of choice so relentlessly,
the more profound freedom is *the freedom not to choose*.

Magic is often defined as the art of producing changes in consciousness—
and thereby in reality—according to will: it's kind of a vague definition,
but people who practice magic have never been known for transparency

Even so, it's hard to tell, by this definition, what *doesn't* count as magic,
starting with getting up in the morning. One minute I was blissfully asleep,
then I found myself dressed and in the bathroom, brushing my teeth!

By what magic did some ensemble of dark forces make a zombie out of me?
Much of our lives are run by magic, dark and otherwise. To count as magic,
there must also be something definitive about *how* the will is exercised.

It might be by some profoundly circuitous path (as is often said *the will of God
works in strange ways*), or at a distance, without ever making direct contact,
often by some obscure resonance (as by sticking pins in voodoo dolls),
or via something (the wave of a wand) tiny in comparison to its effect.

We will be exploring this *how* of magic. And (in the interest of full disclosure)
there is a change *I'm* willing here—some magic I hope to perform in this text.

The word *will* is a bit creepy, tinged by macho melodrama and even fascism
(as in the Nazi film *Triumph of the Will*) with their bloody-but-unbowed heroes
and heroic failures. So part of this project has to involve a redefinition of the *will*.

To start, I like Blake's "firm perswasion" (improved by the charm
of his irregular spelling), related to what he called *enthusiasm*
(in the old sense, from *en-theos*: to be possessed by a god).

> ASIDE: Blake on Firm Perswasion
>
> The Prophets Isaiah and Ezekiel dined with me,
> and I asked them how they dared so roundly to assert.

that God spake to them; and whether they did not think at the time,
that they would be misunderstood, & so be the cause of imposition.

Isaiah answer'd. I saw no God. nor heard any,
in a finite organical perception; but my senses discover'd the infinite
in every thing, and as I was then perswaded. & remain confirm'd;
that the voice of honest indignation is the voice of God,
I cared not for consequences but wrote.

Then I asked: does a firm perswasion that a thing is so, make it so?
He replied. All poets believe that it does, & in ages of imagination
this firm perswasion removed mountains.

The magical will is neither of the diametrically opposed positions
called *belief*: that is, (1) a position based on empirical evidence
and rational argument, or (2) a position based on faith.

Will is related to confidence, which is why a con game is a real form of magic.
Systematic confidence, known as *entitlement*, comes from the very unmagical
and systematic privileging of particular kinds of people over others.

Even so, there is something magical about the kinds of empowerment
it can produce—the spell it can cast on others (at least those who buy into it)
and, following from this, the way it can become a self-fulfilling prophecy.

Look at politicians actually elected: *and you say you don't believe in magic?*

This is also why Tom Paine called aristocratic titles "circles drawn
by the magician's wand." Because they bear meaning only by collective consent,
Paine predicted they'd wither "when society concurs to ridicule them." He was right.

Will is linked to desire but distinct, as I found out when I quit smoking,
which I never really *wanted* to do. Hence the necessity of *willpower*.
Understanding this distinction was exactly what enabled me to do it.

Will is *distinct from* but *comes out of* affect and emotion
(deriving from "the anticipation of a meeting between
the subject's body and another body, real or imaginary").

Parents serve the baby's sense that it is the center of the universe:
it gets hungry, and the breast magically appears; it struggles
to grasp a toy, and the toy is magically placed in its reach.

Thus all exercise of will—and with it, all magic—*comes from having been loved*,
from being granted the fantasy of omnipotence and then being weaned from it
and seduced and tricked and empowered to build actual agency on it.

Accordingly, the magical will is not an executive function—the lightning
that issues from the angry god's outstretched arm—but a circuitous *linkage*
(which is, in fact, also a much better description of how lightning works).

So when I talk about what I'm willing in this book, it is not some manipulation
I'm trying to perform, more like a direction I'm moving; a landmark I glimpse
in the middle distance; a star to which I am hitching my wagon. Simply this:

a better understanding of magic as a category of experience,
linked to related systems and processes in the world at large.

> ASIDE: Metafictional Magic: The Writer's Will
>
> Virginia Woolf's famous book-length essay *A Room of One's Own*,
> adapts lectures she gave in 1928 at Newnham and Girton Colleges,
> the first Cambridge colleges to admit women (starting in 1869).
>
> In it, Woolf develops (so it seems) a simple point:
> for a woman to thrive as an author, she must have
> a modest independent income and a room of her own.
>
> Like a magician introducing a trick, calling on the audience as witnesses,
> Woolf announces upfront that "I am going to develop in your presence
> as fully and freely as I can the train of thought which led me to think this."
>
> In the spirit of showing that she has nothing up her sleeve,
> she reveals that she will do this by putting a fictional persona
> (which she will call "I") through a series of fictional incidents.
>
> These will come to include, for example, the critique
> of a hypothetical book by a hypothetical novelist.

Throughout, Woolf reminds us that it's fiction—
a strategy now widely known as *metafiction*.

Part of the magic is that *we will still believe her*, in spite
of the fictional backstory, because (for a start) her thoughts
are still real and compelling, and her touch is so deft.

Her assertions of fictionality may well incline us to believe instead
that *they are themselves fictional*, and that much of her account
is thinly disguised autobiographical nonfiction. A nice trick!

In addition to inventing a narrator and a novelist, Woolf invents
an imaginary woman from the distant past: Shakespeare's sister,
a genius thwarted at every turn by her social status as a woman.

Woolf returns to this figure at the end of the lectures,
where her magic is to make this fictional figure *present*.

"This poet who never wrote a word and was buried
at the crossroads *still lives* ... in you and in me
and in many other women who are not here tonight,
for they are washing dishes and putting children to bed."

As in her opening strategy, Woolf's trick was to gain power
for this *reveal* by situating her character first as fictional.
Since genius requires "the habit and freedom and the courage to write
exactly what we think," Woolf exhorts women to cultivate these habits
to pave the way for the almost messianic emergence of female genius.

Will there be a female Shakespeare? Woolf ends by asserting
that "she would come if we worked for her, and that so to work,
even in poverty and obscurity, is worth while."

This linkage of a hypothetical past, present, and future
is only understandable as an act of affective labor: the magic
of an individual and collective reshaping of space and time.

We could argue whether there have been women Shakespeares
since then, whether Woolf was one of them, or whether the icon
of singular genius has always been a largely patriarchal myth.

In any case, it is remarkable that *all Woolf asked for* in her essay—all
the specific kinds of world-changing texts she wanted women to write
"if you would please me": *all of these have been and are being written.*

Say that the essay, with the wave of its metafictional wand,
conjured them to be written, not by an executive act of will
but by installing itself in the brains of generations of readers—

by linking itself with other thoughts and feelings and actions,
joining together with them, not by direct causality but circuitously,
like the fluttering butterfly and the tornado.

"I wrote it with ardour and conviction," Woolf noted,
in spite of the parts that are "pitched in too high a voice"
as she put it in her diary, unerringly self-critical: a voice
"too high" suggesting both her feminine persona and elitism.

In spite of this, and in spite of the meandering path the essay takes,
as you read it *"you feel the creature arching its back & galloping on."*

The creature—a greyhound, so it seems—is *her will*, her ardour
and conviction, or less aristocratically, her *doggedness*.

You feel the creature arching its back & galloping on.
Yes, Virginia, we feel it! The magic comes *through* the feeling,
and through what readers empowered by it have done with it.

History isn't made by heroic exhortations
of a single writer; that's Woolf's point. It is made
by women (like Woolf) applying their shoulders to the wheel.

Her efficacy lies in how she waves the wand (the supple cadence
of her sentences), the clarity and intensity of her vision,
how it recruits feelings (hers and ours) for a feminist project.

ASIDE: Shapely Sentences

The kind of authoritative and declarative sentence
"made by men out of their own needs for their own uses"
(as Woolf put it) was one that "Jane Austen looked at

and laughed, 'devising instead' a perfectly natural,
shapely sentence proper for her own use."

A famous shapely sentence begins *Pride and Prejudice*:
"It is a truth universally acknowledged, that a single man
in possession of a good fortune must be in want of a wife."

The sentence operates via the layers of metacognition
it mobilizes: to write or read it properly, to catch the tone
requires being both in and out of the category of "women";
in and out of a patriarchal organization of desire.

Such metacognition, with its neurological complexity
(often called "theory of mind"), is an achievement
underlying abilities such as empathy, self-consciousness,
and lying—in other words, the novelist's basic skillset.

I'll wager that neurologists monitoring brain activity
of people reading Austen and Woolf could verify
that something we can call "*écriture feminine*"
(feminine/feminist writing, as named by theorists)
really does rewire our brains.

In any case, it's easy to see what Woolf—
with her own shapely, metacognitive sentences—
found in Austen, and how one might support her claim
that middleclass women beginning to write novels
was a historical (a *neurohistorical*) turning point.

This is why Woolf is so insistent that anger and reactivity would poison
her vision (a bit too insistent, as a less genteel, modern reader might say).

To be pulled into reactivity is to lose one's own center of gravity—
or simply to lose, by allowing the opponent to set the terms.

Call this book a simple *accounting* or *labeling procedure*:
we're going to go through various promising examples,
establishing definitions and protocols for reorganizing
things according to *what might count as magic*.

ASIDE: Beckett Meets *Twilight Zone*, Glass-Half-Full Version

Curtain rises: Editor and Author are standing stage right.

Editor: I've left you a whole worldful of stuff.
Go through it, put what counts as magic in this box,
and leave the box on my desk when it's full.

Author: How will I know which things count as magic?
Editor: They will have the features you've identified as magical.
Author: But how will I create a list of magical features?
Editor: Just look at the magical things and see what they share.
Author: Okay, but how will I recognize the magical things
when I encounter them if I don't yet have the list of features?

Editor: Can't you recognize magic when you see it? Do your job!

Editor and Author remain standing stage right; Narrator enters from stage left, dressed in a black suit and smoking a cigarette.

Narrator (*addressing the audience*): It's an old dilemma. How can anything get started? How can systems make components when, prior to anything, they need components to exist?

The answer is that *this is the magic*.
Systems and their components
emerge and evolve together.

Inevitably, writing about what constitutes magic must involve a process of recognition and re-cognition, a process of *witnessing and testifying* that's more like a magic act than a legal or religious one.

The recognition and testimony—*yes, this constitutes magic*—imply being present in the moment *and* standing back from it.

By doing this, I am trying to participate in *a wild religion*.
It's wild because it doesn't reliably reproduce itself: I don't expect my readers will be writing more books like this one!

I am not even trying to get you to share my world or belief-system.
I prefer it if *we don't*. I prefer similarity to be found *in difference*
(which is why I've always studied *metaphor: sameness-in-difference*).

My lineage is diasporic; I like being a stranger among strangers.
I've never wanted disciples, so I'm happy that the students
who have found me are those who are allergic to masters.

What I want is to remain here on the margins with you,
carving out a place where we can have rambling but intense
conversations with those who know other things than we do!

CHAPTER 2

Complex Systems in a Nutshell

1 Horror Movie Reboot

This chapter develops a field-specific argument about openness and closure of systems, a necessary detour on the way to a deeper understanding of magic and meaning. I've made the case in the plainest language and most interesting examples I can muster.

The chapter also puts much of my own previous work in a nutshell—
as in this book I have plagiarized my own previous writing, here and there
(because, as Bob Dylan said, "You must leave now, take what you need,
you think will last, but whatever you wish to keep, you better grab it fast").

This section also mashes up my account of systems
with Steve Shaviro's 2014 book *The Universe of Things*,
which takes off from the philosophy of A.N. Whitehead.

While I keep trying to distill my work to its essence,
I keep reproducing the otherness (the *not-mine-ness*),
the fundamental heterogeneity and hybridity of it.

This reminds me of the movie about a scientist who invents a teleportation pod
but doesn't realize that a fly has snuck into it with him. He turns into a fly/human,
and after wreaking various havoc, can only crawl abjectly and beg to be killed.

As you will see, my version isn't a horror movie.
The hybrid monster wins—but that's a good thing,
embodying and marking the direction of aliveness.
You may know the story: it's called *evolution*.

2 Interpositivity

Systems are somewhere between *subjects* and *objects*, and as such,
good for rethinking what were, we thought, unambiguously one
or the other, letting us recognize all as denizens of *betweenspace*.

Of course, outside of language (waves hand, gesturing *over there*),
there would be no distinction between *subjects* and *objects* at all—
but it takes work, inspiration, love to think around the distinction.

You may think language is an added-on layer, icing on the cake, an enhancing
but belated evolutionary development that still is in some sense *inessential*.

But isn't the cerebral cortex (the platform for language) itself
a set of added-on layers? Try removing that and see how you do.

And while we're at it, isn't the *whole brain* a belated add-on, and *evolution itself*
a process by which multi-layered organisms are built on simple metabolic cycles—
cycles that, at bottom (before there was even life), were chemical processes?

And what about all the laws and particles that began to spring into existence
only in the seconds *after* the Big Bang began? The *universe* is inessential!

Each of the new layers have been retrofitted,
wired back through the earlier layers.
The icing isn't just on top; it's between the layers.

Language is often considered to be a tool
manufactured by humans, but this way of thinking
shows how subjects, verbs, and objects mislead us.

As mentioned earlier, growing human brains and language,
along with the growing size and organization of social groups,
have co-evolved, catalyzing each other as they go.

Language may be better conceptualized as a parasite
(a virus, as William Burroughs put it): something that bred in us
and co-evolved with us but didn't exactly come *from* us.

When feeling misanthropic, I could say I'm a race traitor,
less concerned with my fellow humans than with serving
the alien being called language (cue the evil laughter).

Language isn't simply an add-on. It changed the way
we ongoingly make and remake ourselves, our *autopoiesis*.

We have become cyborgs, hybrids. And in the process,
our identities and *even our origins* have changed.

After a new tributary joins, it is no longer the same river.
Now we have another origin. We come from the flesh,
from the earth, no less from each other, and from language.

> ASIDE: Interpositivity
>
> A distinction is often made between a *core* or *essence*
> in which identity is grounded and *differential* identity,
> not grounded in any essence but emerging and evolving
> via relational differences from a range of other things.
>
> All identities are relational, but this doesn't mean we have to choose
> the negativity of pure difference over the positivity of pure essence.
> If you like big words, we could call the third option *interpositivity*.
>
> Essence and relationality are looped together in such a way
> that each is continually subject to the other. Both of them
> can be quite stubborn—and surprisingly flexible as well.
>
> For example, the process of evolution means
> that even origins can change retrospectively
> (and not just via some semantic trick):
>
> say your ecosystem shifts and starts to favor traits
> of one set of your ancestors over another set
> that had been dominant in defining your identity.
>
> Over generations, members of your species
> with formerly definitive traits get weeded out.
> *Who's your daddy now?* Another magic act
> of systems: the retroactive shifting of origins.

3   Becoming a System

Since we cannot simply dispense with language
(especially not here in reading and writing,
where we operate entirely through it),

could we imagine instead a grammar—or a way of thinking
necessarily at odds with the grammar in which it's expressed—
in which there would be *no subjects and objects*?

Philosophy continues to be necessary
because the way language is structured
makes various things counterintuitive.

Consider a humble example: the ways in which, by love and sex,
you may begin to come into a relationship with another person
that can be thoroughgoing enough to be called a *system*.

Intimacy and *jouissance* can be convulsive, and—whether
you go down that road or not—contain at least the potential
to open you up and make you part of a system with someone.

This is part of the danger of sex and part of what makes it sexy,
even if keeping it at arm's length is part of what turns you on.

If, like two galaxies, we get too close, assorted solar systems
or even spiral arms full of solar systems may start to pull apart.

Maybe we end up passing by without much damage (losing only
a few million stars each, say), or maybe we end up orbiting each other
in some wobbly fashion, or merging into a supermassive black hole.

You can say: the closure was never at the level of a *self*
that gets opened up and reconfigured by intimacy. That self
was just a convenient construct, a puppet worked by other forces.

Where's the closure, then? Move "up" to what are often
regarded as higher level phenomena and there's even less
closure to be found in language, ideology, society, culture.

For example, the snapshot of a language may give you the sense
that it's a closed system—capturable in a dictionary, where all words
are circularly defined with reference only to the other words.

Take the long view, though—the hundred-thousand-year view—
and you can observe language emerging from non-language,
just as you would find with flesh-and-blood creatures.

Even in the ten-thousand-year view, you'll see languages
continuously merging and morphing and dissolving
and recombining with each other (just as galaxies do).

More importantly—and even in the ten-year view—
languages enter into shifting ecological relationships
continually with an array of *things that aren't language.*

Move "down" and you won't find closure either. For starters, every cell
of all living things derives from the merger of two or more kinds of creatures
(which is why our mitochondria have different DNA than the rest of us).

Fact is, becoming part of an emergent system is transformative.
What applies to the construct we call *self* seems to be the case
all the way down and up: no absolute closure is to be found.

And by the way, if there weren't *almost* total closure, if we
could merge into happy mush, love and sex wouldn't be sexy,
nor would they be if there were total closure and no danger.

4   Creatures of Light

In complex systems theory, self-making open systems are characterized
by complementary qualities: *operational closure* and *structural coupling.*
It's easy enough to get the basic idea of what is meant by these.

Operational closure refers to the set of processes
cyclically constituting and reconstituting the system,
starting, in the case of a living being, with its *metabolism*.

Systems are usually—to varying degrees—described as having integrity and unity
that come from their closure. It is a provisional and precarious unity. It can change
and evolve, but it's still a unity. Even so, such a system isn't hermetically sealed.

It can be spurred from outside to change, but this can happen
only by how it *metabolizes* the perturbance: *on its own terms*
or *not at all*—even if, in the process, its own terms change.

Hit one rock with a hammer and its molecular structure alters.
Another shatters, or cleaves along certain lines, and another
just shrugs it off and continues its phlegmatic existence.

> ASIDE: Transformative for Whom?
>
> Whose life and identity could be changed by *these* ideas?
> Someone to whom the ideas already mattered very much.
>
> The idea that the earth revolves around the sun was a radical shift
> only for those who had invested a lot in an earth-centered universe.
>
> The question of what stands at the center wasn't definitive
> for Chinese cosmology, so Copernicus didn't rock that world.

Here's another kind of example already touched on: subatomic particles
stream through our bodies constantly without us even noticing.

Although we, ourselves, in the flesh, are mostly insensitive to this subatomic crossfire,
we have created apparatuses exquisitely sensitive to subatomic to-ings and fro-ings.

We've further bound ourselves to them, by electricity, atom bombs, the internet.
But our relationship with *photons* (as with electrons) has *always* defined us.

We are creatures of our co-evolving relationship with light.
We grew up together. We living creatures got *interested* in it early
(developing light-sensitive spots) and adapted ourselves to it.

We have also adapted *it*: done things with light and to it
that the universe hadn't thought of until we came along.

Plants invented photosynthesis, but we humans have expanded
the repertoire of making things with light, such as our new,
complex ways of making and transmitting information.

We selected photons as mattering to us in very particular ways.
Our selection made them matter in these ways, but they had to be
qualified for the job: we both *find* and *invent* what matters to us.

In all this, it is deeply misleading (or let us say instead
that it's a *figure of speech*) to use "we" as the subject:
we are just as much verb and object, process and outcome.

Rather than saying "we selected photons," one might
talk about the universe in the process of self-realization
or photons exploring their capacity to matter in particular ways,
but any way you slice it, *meaning goes all the way down.*

ASIDE: Plant Sorcery

Most mammals, including primates, have *two* cone opsins—
the proteins in our retinas that allow us to see color.

A genetic mutation enabled our ancestors to develop *three*,
allowing us to see almost a million new shades of color—
in particular, those colors along the red/green axis.

Glitches are common, so some of us still get the earlier version,
leading to red/green color blindness, mostly among males
but not females: if one X chromosome has the earlier version,
the second X can still provide the three-cone-opsin update.

The three-cone-opsin mutation stuck, apparently, because
it conferred an evolutionary advantage, especially
for seeing reddish fruits and berries, ripe ones in particular.

Making fruit gathering easier and faster, freeing up time
for getting other foods, allowing us to ingest more calories
(and so develop bigger brains) had big repercussions.

A small thing in itself, the three-cone-opsin mutation helped set off
a kind of "positive spiral" in our development, a feedback loop.

By returning various rewards, it led to the conservation of the mutation
and a growing constellation of further developments that build on it,
making it crucial to what we were and what we have become as a result.

But this story can also be told with another protagonist.
Plants can just as accurately be said to have gone about *cultivating us*
as part of their reproductive systems, making themselves attractive to us
by color, pattern, scent, taste, and sometimes by drugs.

By these means they insured that we would eat them (and sometimes
come to cultivate them in turn), spreading their seeds in the process.
But they didn't just beguile us with caricatured feminine wiles.

They *engineered us*—our vision in particular—by creating
an incentive program that rewarded those of us
who could better see them and thus better serve them.

Those wizardly, devious plants! Making life from sunlight,
water, and dirt was only their first and most famous trick,
but they've kept on expanding their repertoire.

They fine-tuned human vision and everything built around it,
perhaps including humanity itself. What will they think of next?

The rewrite with plants as the heroes is no more fanciful
than the one that put humans at the center. But if you do want
to descend into fancy, consider the following questions.

Can we discern a mythic account of the *humanizing* effects
of better red/green distinction (and thus better fruit-gathering)
in the story of Adam and Eve and the apple? Is it relevant
that the ability to distinguish red/green belongs disproportionately
to women? Or that another adaptive benefit of improved color vision
is the ability to see (can you guess?) especially dangerous *snakes*?

5   Tornados, Whirlpools, and Fires

Operational closure is what enables a system's *autopoiesis,* or *self-making*.

For example, as long as my cells and organs continue to make and remake themselves
in a continuous cycle (this cyclicality *is* the closure, sometimes called *circular causality*),
I stay alive. This isn't a *precondition* for being alive; it's a big part of what being alive *is*.

Our autopoiesis is possible only because we are also *open systems*
structurally coupled with our environment. Stuff has to flow into us
(the stuff we make into ourselves) and stuff has to flow out of us.

In this model, the openness of structural coupling
is the flipside of the system's operational closure.

Like tornados, whirlpools, and fires, we are creatures of self-making, self-sustaining sets of processes—*orchestrations of inflows and outflows*—and in the midst of this dynamism, operational closure allows us to establish entity and identity.

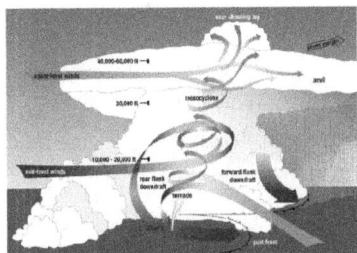

We understand easily that a fire that has been stoked for hours is the same fire, though the fuel and the flames have changed and the fire may even have moved from its original location.

The fire is a continuous thermodynamic difference between flame and air. If it went out and had to be relit, the continuity would be lost. It wouldn't be the same fire.

If it merged with another fire, again you couldn't make a good case that it was still the same fire. The continuity would be maintained, but in the merger, thermodynamic difference would have been lost.

The selective orchestration of flows by a fire or a tornado is its *autopoiesis* (how it makes itself), its *operational closure* (setting it apart from its environment), and its *structural coupling* (binding it to its environment).

As always, grammar is misleading. It's not quite right to say that there is an entity called a tornado or a fire (what we usually make the *subject* of a sentence, a noun) that orchestrates (*verb*) inflows and outflows (*objects*).

The entity *is* the performing of the actions; the performance and actions *are* the entity.

> ASIDE: iii
>
> The reign of subject/verb/object isn't total. When we say *it's raining*, it's less a subject/verb, more an overall event, because we think of it as a general environmental condition.

When we want to distinguish it *from* the environment,
we revert to subject/verb/object using "the rain" as subject.

Following from this, one might try to imagine a language
in which there were no nouns, only participles. Consider
that "I am" is just a calcified, eroded version of "it is i-ing."

This prompts me to consider joining the ranks of those who invent
pronouns for themselves, in this case, *iii,* standing for "it is i-ing."
(This might be the second Yiddishism I've invented in this book.)

What sets a system *apart from* its environment is also what makes it fully *a part of*
its environment. There is a word, one of those words that acts as its own opposite,
that means both *separation from* and *joining together with*: the word is *cleavage*.

Isn't *cleavage* a sexier way of thinking about it?
Unless you prefer the machinic sound of *structural coupling*—
and then again, who doesn't, at least every now and then?

> ASIDE: Co-evolution
>
> The *auto* in *autopoiesis* is misleading in a familiar way:
> it overemphasizes the *self*. Selves can't make themselves.
>
> The *making* or *poeisis* takes priority, a sustained event
> out of which *emerges* a self (foregrounded, at least to itself)
> and an environment (backgrounded, but again only to the self).
>
> *Subject* and *object*, *self* and *environment* are only how things tend to look
> *to the subject itself*. Otherwise, we'd better focus instead on *co-evolution*.
>
> In co-evolving, selves and environments are continually at stake
> and in process of being made—constellated by the wiring together
> and the retrofitting of systems, subsystems and supersystems.
>
> Think of this shift not as categorical but simply as a change in emphasis.
>
> Even so, it can help correct the way we *overestimate* ourselves—
> as the foregrounded heroes of the story. And just as crucially,
> it can help correct the way we *underestimate* ourselves when we fail
> to recognize the extent to which our worlds are *of our own making*.

In the process, we stand a much better chance of finding
and coming to terms with what freedoms we might gain
and what fates we must embrace (or suffer the consequences).

In the process, maybe we can also learn to skirt the tragedies
that stem from overestimation—*hubris,* as it is often called—
and the tragedies that continually unfold from underestimation.

These latter seem to derive from something passive-agressive in us—
an *othering* whereby we disavow or ventriloquize our own agency.

6   Leveled

As we're seeing, complex systems theory relies on binary distinctions:
operational vs. structural, closure vs. coupling, system vs. environment,
self-sustaining and self-reproduction (autopoiesis) vs. annhilation.

Press on any of these distinctions—as we've started to do—and they tend to collapse.

This doesn't just invalidate; in fact (as we will see),
it simply shows how complex systems theory
performs, sustains, validates and *reproduces* itself.

You could spin this negatively by calling it a narrow, self-fulfilling prophecy,
and/or you could maintain that this is exactly what *all* successful systems do.

My point will be, instead, that we have a great deal to lose by failing
to acknowledge the central paradoxes as such, as well as a lot to gain
by acknowledging them as such and putting them into practice.

We've acknowledged the paradox that what sets a system
apart from its environment is also what makes it fully,
and *even more intensely,* a part of its environment.

Its boundaries both *prohibit* things from flowing into
and out of it and *mandate* such a flow—*selectively*.

This selection is like the thermodynamic difference made by Maxwell's Demon,
that imaginary creature who sits at the gateway in a closed, bifurcated box,
letting the fast particles go to one side and keeping the slow ones on the other.

Systems manufacture dynamic difference.

But as noted, far-from-equilibrium systems don't emerge *from* equilibrium. Nothing can emerge without instability. No pearl without a grain of sand. No creation without symmetry-breaking. We are not watchful demons in hermetically sealed boxes at thermodynamic equilibrium, making something from nothing. Sunlight streams in against the cold and dark, powering a cosmic heat engine. We surf that thermodynamic gradient, and we are thermodynamic gradients in turn, waves on waves.

Structural coupling is less like a steel beam welded to
a steel framework, more like a smaller eddy in a whirlpool:
a disturbance that the system conserves and elaborates.

But (as another homely example will show), though we are waves-on-waves, it isn't just a fractal regress, a hierarchized set of levels.

Let's say I suffered a traumatic brain injury and went into a persistent vegetative state. You might still say it's me—*sort of*—like the smoldering of what had been a bonfire.

> Is that him over there? No, but *that autopoiesis
> was his*. You mean *he was its*, don't you?
> Well, it was a bit of both. But alas, no longer.

If left to my own devices, I'd die quickly,
no less than an unattended infant would,
or as I would if cast adrift in outer space.

My body could continue remaking itself physiologically,
but because I couldn't provide for myself, my structural
coupling with my environment would be compromised.

However, since I'm a social creature, I could still be kept alive—
not necessarily by extreme technical means but by simple human care—
perhaps for years. (Although, just to be clear, that isn't what I'd wish.)

> ASIDE: On being born too early and dying too late
>
> Humans are inclined to take care of those who'd otherwise die. Starting with offspring who need extended care after being born, it's an evolved characteristic. It's why *letting go* is so hard for us.

Though vegetative, I would remain *some kind of player*
in the web of ecological, social, economic, and legal relations
in which I once moved. True, I would be less of a player
and more of a gamepiece, but as always, it's a bit of both,
and difficult (or even heartbreaking) to sort them out.

Of course, provisional distinctions can be made
among *a body, a mind, a society,* and *an ecology,*
but this ensemble keeps me alive and keeps me me.

They aren't numerable layers or concentric circles but a woven-together mesh
of heterogeneous strands, a *tangled hierarchy*. This is why Francisco Varela
called an organism a "meshwork of selfless selves": less and more than a self—
more plural but still a kind of unity; more *federation* or *syndicate* than *nation*.

Note that I apply qualifiers—"a kind of unity" and "more plural"—
to what are normally defined as unqualifiable: supposedly,
something either is a unity or not, is either plural or singular,
so there shouldn't be *kinds* of unity, or *more or less* plurality.

I have called *someness* the condition of being between singular and plural.
What if there were no unity and plurality, only kinds of someness?

Again, this might not be so counterintuitive if our grammar
didn't insist so often that we choose between singular and plural.

> ASIDE: Up Around the Bend
>
> I sometimes think of paradoxical difference (as between singular
> and plural as varieties of someness) via the proposition set forth
> in the old Creedence Clearwater song "Up Around the Bend":
> There's a very different world up there, and although
> you can't see or hear it from here, it is not entirely separate.
>
> In the song, the other world is very like what Northrop Frye
> called the "Green World" of pastoral romance, or what Bakhtin
> called "Carnival," a mirror-world—e.g., art set apart from life.
>
> You may think of this as simply another way of saying
> that, beyond a certain tipping point, *difference in degree*
> can become *difference in kind*, but it goes beyond that.

There isn't a sharp corner or a point at all, and paradoxically, this makes all the *more* difference between the two worlds.

Up-Around-the-Bendedness keeps the difference from being contained and domesticated by a dividing line, putting *more* radical heterogeneity and otherness into *the same space*, cabbages, kings and the kitchen sink.

7   A Personifying Universe of Stretchy Things

The paradox at the heart of the opposition between self-reproduction and annhilation yields important insights.

Rather than saying that complex open systems are driven to perpetuate themselves and retain their integrity, it is just as accurate to say we surf on our own dissolution.

The reason a wave develops a surfable structure is the same reason it must break: because it is in the *process* of breaking. This isn't just a metaphor of our situation as living beings; it marks the virtual kinship that waves and living things bear.

Systems that perpetuate themselves and police their boundaries—
as they are often said to do—sound a lot like *nations* and *empires*.
There are, you may have noticed, steep costs for seeing things this way.

You can say that both perspectives are valid—policing order, surfing dissolution—
and that both must always be in play, but notice how much the emphasis matters.
And notice it's the surfer, not the policeman, who says that "both are valid," dude!

If the perpetuating/policing aspect seems less dominant to me,
maybe it's because, like you, I'm a diasporic and exogamous person.

Moving among strangers, I don't fear that my ethnic, class and culture mix
will simply disappear into an ever-homogenizing general population.
I assume that my own mixing will contribute to new sets of differences.

I feel the same way as a teacher: as I've said, I don't want to produce disciples,
a mini-me to carry forward my ideas. Instead, I have sought out and been sought out
by those with whom there's enough rapport for some degree of mutual transformation.

This may also have something to do with being a non-breeder.
It is also the way I feel about my partner and her children.

It seems to me that our love has transformed each of us,
but it wouldn't capture it to say it's made us more like each other.

I like to think we've been catalysts for each other's evolution,
and though there is reciprocity, there is also radical assymetry
in terms of what we've gotten from each other. Isn't a gift
something *not* repaid in kind? *Isn't this what love is?*

You may think I've now put this too personally—
in too much of a touchy-feely, New-Agey way—
but *my argument mandates that I do so.*

I've let it stray into the mode of the "personal" and into the realm
of "feelings or preferences or principles that constitute my identity"
because this is my style and because I am arguing that *style is identity.*

This is how I practice what I preach: *to the choir.*
It's how I enable others to recognize themselves in this
(and other others to *disidentify* from it and from me).

It is a way of enacting what Steve Shaviro, developing the philosophy of Whitehead,
identifies as the precedence of *the aesthetic*, which involves preferences—*selections*—
for some things rather than other things. Either this invalidates ontology altogether, or
(my preference) the precedence runs deep enough to qualify as an ontological claim.

This is why I spoke of our preference for photons in the building of our identities
(long before there was even an *us*) and implied that we could also retell that story
in terms of choices photons make in their process of exploration and self-realization.

It stretches things to say *choices* and *self* when talking about photons,
but *this is precisely the point.* It stretches things in exactly the same way—
only a bit more—as it does when we apply those same words *to ourselves*—
as if we were sovereign entities exercising our own godlike agency.

Personification is a figure of speech, a way of stretching things.

Yet personification is not just something that happens in language,
not just the way our personifying brains make sense out of otherwise
impersonal constellations of forces. The universe generates entities.
(Warning: ontological claim!) *It's a personifying universe of stretchy things.*

8   Dynamism

Opposing *autopoiesis* and *annhilation* obscures
what might be a Daoist or Buddhist principle:
the process that leads to fragmentation and dissolution
is *the same process* that leads to emergence,
consolidation, sustaining, and elaboration.

Like fire, a life is "consumed with that which it was nourished by."

This is the same idea as "that which does not kill me makes me stronger,"
at least insofar as it rejects the dichotomy between things that *enhance*
and things that *detract from* a system's maintaining itself.

The continual process of reconstituting myself is both why I am so strong
(how I can withstand shocks and heal, like a spinning top that can be knocked
and right itself) and why my life is precarious and can only be sustained for so long.

Even species of micro-organisms (some of which are not subject to aging
and can clone themselves indefinitely) eventually are killed by accidents,
die off by environmental change, or cease to exist by evolving into other species.

Even beyond this, life itself seems necessarily to be more fleeting—
it began later and will end sooner—than the rest of the universe,
which is why I'm (note to self: idea for song) *stuck in the middle with you.*

In case you hadn't heard, the universe is a dynamic entity
whose dynamism is why it can't last forever, either.

In our more recent evolutionary line, *sexual reproduction* (which optimizes the difference
of offspring from parents) evolved along with *programmed death* at the cellular level
(cells continually dying and being replaced) and at the organism level (via built-in aging).

Evolution favored sexuality and mortality because they enable increased evolvability.
Sexual reproduction is an oxymoron. Sexuality was selected as difference-optimizing:
in other words, it was selected because it really isn't *reproduction* at all.

It is misleading to talk about systems reproducing themselves
as if this were the norm and then about mutations as aberrations:
the capacity for mutation is the *why* and *how* of our evolution.

ASIDE: One Plus One Equals Infinity

To get slightly more technical about it, what's misleading
about the reproduction/mutation paradigm is its tendency
to foreground difference against a background of sameness.

This mistake may be so fundamental as to be almost uncorrectible,
at least in capitalist modernity, with its valorizing of productivity
(dramatized as *disruption*) and its devaluing of mere reproduction.

To begin, at least, to address the mistake means recognizing
that difference and sameness don't have to be thought
as always-already-present components that *pattern* organizes.

Instead, like a system that produces its own components,
pattern produces difference and sameness, or to leverage it
from the question-begging organization of subject/verb/object,
pattern, difference, and sameness necessarily co-emerge.

If you have (let's say) just two hydrogen atoms,
you've already got sameness *and* difference—
and pattern. Just keep folding and unfolding
for a while, and *voilà*, there's your universe.

You'd think that increased evolvability might lead, in the long run,
to more sustainability for life in general, but the jury's still out.
It could also lead to more intensity at the *expense* of sustainability.

When it comes to planetary murder/suicide, we're the most likely culprits.
Then again, we may also be the most likely ones to send emissaries
off this rock before the whole thing blows. I don't pretend to know. Stay tuned!

9   Magic, but No Black Boxes

It seems that the notion of *representation* goes along with the absolute closure model.
Supposedly, representations differ from things: they are *second-order* phenomena.

I refer to representations not necessarily as correlated with an external world
(that's the naive notion of representation, known as *correlationism*)
but as an organism's way of mediating engagement with its lived world.

You're not looking out a window at a world, but in a windowless room
watching a TV monitor that you think is a live feed from outside
but is more like a simulation recreated in a studio. Pretty grim.

Samuel Beckett meets *Twilight Zone* again,
this time definitely the glass-half-empty version.

The grimness explains the romance of what's called the "great outdoors"
in the philosophical stance known as Object-Oriented Ontology.

If you thought you'd been trapped inside a box (where you could only make claims
about *how you think you know* the outside world but no valid claims about its reality),
then you are bound to get a bit giddy when you find you can step outside.
I grew up in Minnesota, so I know about this. It's called *cabin fever*.

The representation model was a corrective that led to all kinds of insights.
But if the fear of falling back into a naive position (e.g., correlationalism)
keeps you in a reactive mode, you stay tethered to what you oppose.

Representation lends itself to a visual metaphor, which, as usual, brings baggage.

We do not speak of representation by creatures with light-sensitive spots but no eyes,
nor by sunflowers that turn to the light, nor do we call what a sunflower does *seeing*.

We tend to consider the way a sunflower processes light, instead,
as a mechanical chain reaction that never disappears into a black box
where one could find a representation flickering in the dark, closed-off
interior of consciousness: the old movie-theater model of a mind.

In this model, the sunflower makes no representations
or observations, has no mental model of the world.

The categorical distinction makes it hard for people to understand
how eyes could have evolved from light-sensitive spots, even though
the notion that the spots evolved with the creatures—both becoming
more and more elaborate together—shouldn't be so counterintuitive.

Representation is *another modality among many* by which systems
are both *a part of* and *apart from* their worlds. In other words,
we are only *up around the bend* from sunflowers after all.

The rejection of exceptionalism on behalf of what we have called *mind*—
what we have flattered ourselves belongs to us alone—means coming to grips
with how mind goes "all the way down" to subatomic particles and such.

This is called *panpsychism*. Pushing in the other direction, partially dismantling
the black box of mind in favor of the Rube Goldberg Model, we discover all kinds
of gratuitous complications and foldings and detours—*magic*—*but no black boxes*.

> ASIDE: The T-Shirt Version
>
> Wow, can it be I just said everything I wanted to say?
> I didn't know if I could, but it's all in that last paragraph,
> maybe in the five-word slogan *magic, but no black boxes*.
>
> I notice that the slogan is five words, each with five letters—
> well, with Words Two and Three *adding up to* five letters.
> (See what just happened? Two and Three add up to five?)
>
> I finally stuffed it all in and pushed down the lid.
> Maybe I should pare everything else away.
> (What happened to the book you were writing?
> Oh, that. *I condensed it into five words.*)

10   Wildness

We have been talking about *second-order systems*: systems that observe and represent.

For observing systems, as Cary Wolfe summarizes sociologist Niklas Luhmann's model, "all observations are constructed atop a constitutive distinction that is paradoxical."

"The observing system which utilizes the distinction
cannot acknowledge [it] as paradoxical and at the same time
engage in self-reproduction." This is the system's *blind spot*.

But wait, here's your problem in a nutshell!

For the system known as *complex systems theory*, the distinction
between operational closure and structural coupling (along with
related distinctions considered above) must be one of those
constitutive distinctions that can't be acknowledged as paradoxes.

If the theory did acknowledge them as paradoxes, it couldn't reproduce itself—or at least (I want to say) *that's the fear*.

> ASIDE: Pet Resemblance via Social Theory
>
> The kind of buffering this requires (to keep the contradiction from bringing down the system) enacts a (sometimes infuriating) kind of closure.
>
> I've disagreed with Luhmannians before, and what they say is a mix of several responses: "some things you got wrong; and at other points you have mischaracterized our position."
>
> "In some cases we really are in agreement, and in other cases your objections have been anticipated and incorporated by the theory in advance. If you would just read it more carefully." *Seamless!*
>
> You have to give them credit for recursive consistency: the theory resembles the systems it describes (*excessively closed*), as an owner comes to resemble his pet.
>
> Or is it the other way around? A bit of both, no doubt. And, I should add, I hope you can excuse my cattiness!

What if we could *partially* acknowledge the central paradox that "A equals B" and, at the same time, "A does not equal B"— that is, that closure and coupling are both the same and distinct? And what if we can *partially* operationalize this paradox?

What if there were a radical middleground between *reproducing* and *failing to reproduce*? In other words, what if reproduction *and* the inability or failure to reproduce were both only *partial*?

Isn't this exactly what we've already seen with sexual reproduction? The offspring differ, so it isn't really *reproduction*—not as cloning is. Mutation must be fundamental for evolution to be possible at all.

This is *wildness*.

Plant the seeds of an apple and you'll grow trees that bear very different fruits: some will be small, some large, some sour, some (if you're very lucky) sweet, and, in the fullness of evolutionary time, some will no longer be apples at all.

I am not saying there can be a system without blind spots.
It's worse and better than this: the plurality is more radical
than "a dialectic of blindness and insight" could capture.

Blind spots are located where a patchworked organism—
a non-self-identical organism characterized by *someness*,
a "meshwork of selfless selves"—is joined to itself.

We aren't just big eyeballs. The fruits of second-order observations
can sometimes be tasted, smelled, or touched even when not seen
or understood, and whether or not we can fully metabolize them.

Even where *witness* is impossible or abjured,
where language asserts its closure from the world,
even here, *withness* is borne, embodied, acted out.

Even where we assert exceptional difference, our unique ability
to make meaning in a meaningless universe, we are most akin
to our grandparents, wizardly plants who make life out of non-life.

They love and indulge us so much that, even as they continue to tinker with us,
they let us think we are the chosen and choosing ones, the apples of their eyes,
perhaps even up to the point when they decide that their experiment has failed.

CHAPTER 3

Magic by Example

1   Failed Magic: Modernist Heaven (and Hell)

A man and woman are looking out
at something happening in front of them,
shining brilliantly and illuminating their faces.

They smile beatifically. She is holding onto him,
as one would hold onto someone one loves
in a moment of cosmically historic magnitude.

> Surely some revelation is at hand;
> Surely the Second Coming is at hand.

To their right, we see the vision that has so enthralled them.
It's a modernist paradise of soaring, sterile spaces of metal, glass,
and soft bluish and pinkish lighting, in which a few individuals
and groups, of two or three people each, chat or stroll about.

In their midst there might be one lone person of color, a tiny figure
who seems to be alone, frozen in place, and missing a hand and a face.

Each person or small group is allotted something like a hundred square meters of personal floor space (because modernists, apparently, like personal space).

Next to the shiny, happy couple are the words "We're working hard to make Heathrow an airport London can be proud of." This claim is, frankly, a bit deflating. Not exactly the arrival of a messiah!

In addition to suggesting that (1) one should set aside annoyance at the construction (in the name of civic pride and solidarity with the workers), the tagline telegraphs that (2) to Londoners, currently, their airport is not exactly a source of pride.

And since "we're working hard" is presumably *the best* that can be said, (3) there is little reason to hope that the goal might actually be achieved.

How could the designers of this campaign have been so unaware of the unsaid-but-loud-and-clear messages they were sending? Wasn't it their job to cultivate and deploy such awareness? Can we chalk it up to a *failure of empathy* with their audiences?

The campaign invokes a modernist notion of civic unity
in which all of us have a stake, all are working hard,
all share the pride and identify with the same shining ideal.

Didn't anyone point out that we live in a postmodern world
of rising and savage inequalities, and that most of us—
99% or so—don't buy into this "we" anymore?

Rather than identify with the shiny couple, we imagine marketing
meetings where slogans and images are approved, mind maps
and psychological triggers identified, focus groups scheduled.

How did the magic fail? Think of the ad as a kind of self-organizing system:
it emerges as a nodule of meaning, but (as with most kinds of order) at the cost
of creating disorder around it: a neat little message amid its noisy echoes.

You might imagine a tightly spinning tornado leaving a swath of debris in its wake.
But as we've seen, the remarkable thing about such systems is not the instability
that makes them fleeting but the sustainability that allows them to exist at all.

Even over their relatively brief lifetimes, they manage to orchestrate titanic downdrafts,
updrafts, inflows, and outflows so as to keep themselves going. (Jupiter's red spot
is a huge, spinning storm that has been raging for at least 300 years.)

The ad, on the other hand, is immediately overwhelmed by its own back-lash,
its noisy echoes. If it fails, it fails because it fails to become part of a system.

Successful magic may be found or applied in the emergence of systems,
at the acupressure point where the system loops back to connect both
with itself and the world, the juncture where world and system co-evolve.

The Heathrow campaign cast its spin with modernist ideology
that, like Jupiter's Red Spot, has been spinning for 300 years.
Now, however, modernity is wobbling and weaving.
Its magic is awash in its own toxic byproducts and backlashes.

In contrast to the Heathrow campaign, the signature strategy
of many *post*modern ads is the ironic delivery of the message,
folding in the countermessages upfront to neutralize them.

You could call it a dialectical maneuver, or just a magic trick that works (as so many tricks do) by *misdirection*. Irony isn't subversive, or rather, it works to undermine an ideology along with *all that might oppose it*.

In the process, it works to replace the straight version of the ideology with its nihilistically undead double, which may be harder to defeat: "Haha, of course we insiders don't believe in it, but it *is* reality."

So, if the cynics who made the ad had used it to wink at us ironically instead of treating us as dopes and dupes—if they had simply dialed up the kitsch a little in the *retrofuturism* direction—might it have worked?

Maybe. In any case, you will not be surprised to learn that the construction did not achieve a modernist utopia. I came through after it had been done. Everyone stands and mills around in a central hall packed with people.

Departure gates are announced just before boarding, so everyone anxiously watches the big board, and as soon as gate numbers appear on the board, groups of people dash off to the gates.

It's hard not to imagine Heathrow employees ensconced in a control booth, sadistically deferring gate announcements as they enjoy the rising anxiety, cackling as they finally display the awaited data and watch the people dash.

Of course, the failed ad campaign is a trivial example of counterproductive modernity. More epic counterproductivities include climate change, plagues, ecocatastrophes of the Anthropocene Era (in process), fascisms and perpetual wars, cycles of violence, dehumanizing alienations, exploitations, inequalities, oppressions, pathologizations.

It's not clear how long the system known as capitalist modernity and postmodernity can keep orchestrating these violent downdrafts and outflows to keep on spinning.

But modernity also spins off *positive* byproducts.

In Bruno Latour's account, modernity's insistence on separating nature and culture is a sustained trick of misdirection that has operated instead as an engine for *generating nature/culture hybrids* under our very noses.

Hybrids are all around: proliferating and interlocking constellations of humans, technology, animals and plants, soil and air and water. Our worlds are, upon closer examination, all nature/culture hybrids.

Consider the internet, proliferating at new scales of subatomic smallness
and global largeness: quintillions of human-orchestrated electrons and photons
now being marched around the planet, through cables and through the air.

As with other creatures and ecosystems, we cannot understand it in itself
but only as a kind of nervous system emerging from and being retrofitted—
rewired back—through all the other layers of interlocking constellations.

The magic of modernity has been to sustain itself, for a few centuries.
If you haven't yet been compelled to recognize the collateral damage
(where have you been?), it's hard not to hear the whole thing creaking.

Sometimes in the most trivial things, *the canaries in the coalmine*,
we can catch a glimpse of the whole system's fate. As Blake put it,
"The dog starv'd at his master's gate / Predicts the ruin of the state."

"We have never been modern." The myth of modernity was that it had no myth.
We are beginning to see that modernity never was a disenchantment of the world
but a re-enchantment on new grounds. We can begin to see now that it is waning:
"the owl of Minerva takes its flight only when the shades of night are gathering."

We are not flying toward the light—death or enlightenment—nor falling back,
slipping into darkness again, but remaining, at the leading edge of aliveness,
in a kind of Twilight Zone, finding ourselves in the middle of a dusky wood.

2   Placebo Magic

In a Harvard Medical School study, patients
suffering from Irritable Bowel Syndrome (IBS)
were told they would be taking placebo pills.

The pills, they were told, were "made of an inert substance, like sugar pills,"
and that placebos "have been shown ... to produce significant improvement
in IBS symptoms through mind-body self-healing processes." They were told
"that they didn't have to even believe in the placebo effect. *Just take the pills.*"

In spite of being told upfront they were getting placebos,
"almost twice as many patients treated with placebo showed
adequate relief of symptoms compared with the control group."

Even more surprising, "patients taking the placebo
also doubled their average rate of improvement to that
achieved with the most powerful IBS medications."

The lead researcher summed it up as follows: "These findings
suggest that rather than mere positive thinking, there may be
significant benefit to the very performance of medical ritual."

The researcher makes a familiar distinction between *belief* ("positive thinking")
and *practice* ("performance, ritual"). This distinction was implied in the suggestion
that patients didn't "have to even believe in the placebo effect. Just take the pills."

The funny thing is that this assertion—that you don't even have to believe—
might dramatically *increase* the subject's sense of placebo effectiveness
simply by the suggestion that it can work whether you believe in it or not.

The assertion is allowed to do its work all the better
by *relieving patients of the burden of believing or not*,
even to the point of qualifying as *a kind of magic*.

It's a kind of hypnotic suggestion: it only travels *through* consciousness
(since one must at least understand the words for them to have an effect),
being passed along to deeper parts of the brain where it does its work.

Although it's hard to evaluate without having seen the interactions,
the even funnier thing is that the researchers seem not to have noticed
that they are, in some important sense, hypnotizing their patients.

To put it more carefully, the doctors don't seem fully to understand their role
in the ritual even as they perform it, and the *magic* is the part they're missing.

If improvement were through medical ritual, why did placebo patients do *better*
than those who were given actual IBS medications, who presumably went through
the usual version of the same ritual? (I will return to answer this question below.)

It is easy to describe the general way the magic must work. It is uncontroversial
that the parts of the brain that process language (that receive the instructions)
are wired to other parts of the brain (that regulate various chemical balances).

It is also clear that the brain and bowel are wired together in complex ways
and *in both directions*. In fact, the bowel also has its own brain:
the hundred million neurons that make up the *enteric nervous system*.

When you consider that the enteric nervous system is wired
to the brain in the head, is it surprising that this complex ecology
is subject to subtle but powerful interventions and rebalancings?

I cannot make my bowels well by *believing* or *thinking* them well.
This is exactly what is called *magic thinking*—and as with *belief*,
the problem with magic thinking is *not the magic but the thinking*,
at least insofar as it refers to a conscious and autonomous activity.

> ASIDE: Pregnancy via Magic
>
> Someone I know was hoping to get pregnant via artificial insemination.
> She thought that if she *believed* thoroughly enough it would happen,
> it would indeed happen: a classic example of *magic thinking*.
>
> This stressed her out as she tried to police her own feelings and thoughts
> to stay positive. The tight little message (the enforced positivity,
> the *staying on message*) generates counterproductive stress all around it.
>
> On her next attempt, some months later, she got it right, realizing
> it was an exercise of *will* (not of *thinking* or *belief*), subordinating
> all to the task at hand, but *lightly*, with a kind of relaxed stoicism.
> If she had been religious, she might have said, *If it be Your will*.
>
> How much did her relaxed focus have to do with getting pregnant?
> Hard to say, but we know that stress affects such probabilities.
>
> Stress happens when the elements of systems are connected and at odds,
> even at several removes. The elements often include *potential futures*.
>
> Stress is ratcheted up when radically different futures depend upon,
> diverge from, and *redound back on* a present (in which, for example,
> depending on how this goes, one will either be childless or a parent).
> Counter to stress is relaxation and acceptance (on which, more below).

ASIDE: Following the Scent

I thought of the pregnancy story three years later, when I performed
some rudimentary magic: after a long day of play along a river
and in a huge field next to it (covering many square blocks),
we found that one of her kids (she had twins!) had lost her pacifier.

Nobody had noticed when or where she'd dropped the pacifier, but—
in spite of the vast area involved—I felt sure that I could find it.
And sure enough, after walking about a hundred yards into the field
in scrubby grass, along what might have been a random trajectory, I did.
The funny thing was, when I saw a light-colored object
several yards away, I didn't believe it could be the pacifier,
but I went over to look at it anyway—and there it was.

It is a subtle but revealing detail: I simply did not *believe*
it could be what I was looking for. What had happened
(I realized later) was that I *noticed my absence of belief*.

Now as I write this, I notice the same about writing:
it is not that I *believe* that things will fall together,
that answers will come from unexpected places.

I don't believe each detour will turn out to have been The Path.
I just keep my nose to the ground and keep following the scent.

The magic is in *bypassing belief and nonbelief,*
enlisting consciousness to deliver the message
and, without further ado, *getting out of the way*.

Of course, this is not news to athletes and other body-workers. As it is said
by practitioners of Alexander Technique (a mind-body realignment practice),
your consciousness need only put the letter in the mailbox, not hand-deliver it.

You can learn consciously to interrupt and inhibit bodily stresses
and clenchings, but you cannot executively *direct this* to happen.

Magic is the art of causing changes in consciousness and in reality at large
in accordance with will (and, I would add, *via meaning*) but not executively.

Meaning refers to the way one level of a system is connected to another level or one system is connected to another at more than one remove: the way the relationships among elements of a system (its interior ecology) affect and are affected by its exterior ecology; the system/environment interplay.

To describe this process, we must first recognize the plurality and heterogeneity of the brain—a set of Rube Goldberg sequences wired to other such sequences in ways that cannot be domesticated by any categorical distinction of inside/outside.

Meaning operates by triggering rather than by mechanical impact, by repercussion rather than percussion, by *transduction*—the way a system translates an input into another medium; by *transmediation*.

Photons hitting the retina are transformed into electro-chemical patterns, since the brain traffics in such patterns. Brain rewiring, world-changing in turn, derives from how the brain is wired to itself and to the world in both directions.

To end this section with another—happier—Rube Goldberg sequence: "A mother's laughter can even improve the quality of her breast milk, making it more effective at fighting skin allergies in newborn babies."

What makes this sequence sublime (or, on the other hand, a way for scientists to blame mothers for not being happy enough) is that an otherwise linear series of knock-on effects is looped into a system.

The system is wired so that the mother's and baby's psychosomatic well-being are elaborately and sensitively interdependent. Let's say that you're a mother and that you've laughed at some of the stupid jokes in this text, and let's say that you're reading it *p.m.a.* (*post mortem auctoris,* after the author's death).

Let's say that this subtle laughter, like a ninja chiropractor, makes some tiny adjustments that lighten and relax you, and that your baby reaps the benefits of this shift as well. There! I just healed your baby from beyond the grave!

3   Mirror Magic

A tabletop-sized box created by neurologist V.S. Ramachandran is constructed so that a patient missing a limb—say, an arm—

can insert the other arm into a hole in the side of the box.
When the patient looks down through the open top of the box, a mirror
makes it look like the patient has *two arms* (one the mirror image of the other)
and can move them in tandem, as by clenching and unclenching both fists.

This has proven to be an effective treatment for much phantom limb pain.
In some cases, sufferers have gotten relief after years of debilitating pain
and after assorted unsuccessful surgeries and other treatments.

Like magic! But again, is it so surprising that simply *looking* can rewire the brain?
Isn't the brain shaped by its processing of sensory information to begin with?
The mode of the magic is not only in how the brain is rewired by looking,
but in the wiring *in both directions* between brain and body and—here's the key—
in how these are mediated by a third *phantom entity*: a *virtual* or *visual prosthesis*.

Patients don't *believe* they suddenly have two arms.
Of course they are *aware* of what they're seeing.
Consciousness is involved, though there are no words
to process, and, as with the placebo pills, no belief.

The simplest thing one could say is that patients *feel*
what it would be like to have two arms; that they can,
at some neurological level, actually *inhabit* that feeling,

just as an actor doesn't pretend to have a feeling but *has* it.
Words like *feeling* and *imagination* are impressionistic,
but they point the way to a more rigorous understanding.

Surgeries to address phantom limb pain have often targeted
the nerve endings remaining in the stump of the lost limb,
or parts of the brain thought to process the signals they send.

The basic error of such surgeries (as revealed by the mirror box)
is that—as patients have been saying all along—the pain is not
in the nerves or brain but *in the phantom limb*. How can this be?

A closely related parlor trick provides another clue. In this magic act,
you place both of your hands on the table, one slightly off to the side.
A makeshift screen is put in front of your offset hand, so you can't see it.

Your fellow experimenter also places a life-size rubber hand on the table,
and slowly strokes it with a soft brush while, out of your line of sight,
hidden by the screen, stroking your other hand with the same kind of brush.

If the trick works (as it only occasionally does, in my experience), as you feel
your hand being stroked out of sight and watch the rubber hand being stroked,
you start to feel—in a creepy, uncanny way—that the rubber hand is *your hand*.

Our nervous systems can get wired, at one or more removes,
to a detached prosthesis (in the mirror box, a visual prosthesis).

A related case is known as *phantom vibration syndrome*.
Since the rise of cellphones, many people experience
random stimuli (the brush of a pantleg, a nerve twinge
or muscle twitch) as if a cellphone had begun to vibrate.

A prosthesis can be experienced as part of the body,
and the body can be experienced as prosthetic,
and neither of these feelings are simply *mistakes*.

The most famous moment in the most famous role by the most famous actor
of the 18th century, David Garrick, was the one where Hamlet's friend Horatio
shouts, "Look, my lord, it comes," and Hamlet first sees his father's ghost.

Though the ghost had generally been played by a person in a costume,
Garrick preferred to have his own horrified reaction make it seem
as if a spookily invisible presence were there on stage with him.

He was known for introducing a new level of naturalism in acting,
and since acting has continued to evolve mostly in this direction,
we would now almost certainly regard Garrick's acting style
as exaggerated and melodramatic (as did some in his own day).

Even so, love or hate him, all agreed that, seeing his reaction,
you couldn't help but feel he was seeing an actual ghost.

"It made my flesh creep," said the German scientist G.C. Lichtenberg—
or as a journalist put it, just as "no Writer in any Age *penned* a Ghost
like Shakespeare, so, in our Time, no Actor ever *saw* a Ghost like Garrick."

In fact (*spoiler alert!*), a wigmaker had made Garrick a hydraulically operated wig
that enabled him to make his hair stand up. This rather *deflating* bit of information
may unmask his virtuoso acting as a vulgar special effect, but the effect was *the same*.

Horripilation (the silly word for flesh-crawling, hair-standing-on-end terror)
is not under voluntary control, so even great actors cannot reliably access it
through the power of their acting, hence the necessity of the *fright wig*.

Our brains—long before they were even *our* brains—were wired
to detect the presence of other living entities and to induce goosebumps
when triggered by an especially threatening presence.

It's not hard to guess roughly how this could have happened. If a creature
can detect a potential predator or competitor, and looks larger with raised hair,
it is less likely to be defeated or eaten, hence more likely to have offspring.

But Garrick's trick also depends on another bit of brain wiring:
our so-called *mirror neurons* fire when we are performing an action
or, just the same, when we are watching someone else perform it.

Again, this is not under voluntary control, though it is part of how,
via empathy and imitation, we build up our own repertoires
of gestures, expressions, social responses, and even language:
part of how our brains are wired to others and themselves.

Garrick's magic was in triggering a Rube Goldberg set of interconnected mechanisms.

We know (not necessarily consciously) that horripilation is not under voluntary control, so when we see someone's hair standing on end—in conjunction with a horrified gaze—we infer, without having to think, that the person must *actually* be seeing a scary entity.

This inference happens via the triggering of assorted interconnected brain subsystems.

The horror felt on seeing Garrick is just as real a neurological event, though we *also* have the meta-awareness that we are only watching a play. Our goosebumps are just as real—*the same*—as if there were a real ghost.

But there is still another layer or two of magic at work here. Garrick was working up something already *systematized* in the play.

Hamlet goes on to confirm what his father's ghost tells him— that his father was killed by the current King in order to steal the throne. Hamlet confirms this scientifically, by performing an experiment.

He puts on a play ("The Mousetrap") depicting a version of the murder, and as he'd hoped, it triggers the King to betray himself by his response: he is "frighted by false fire," as Hamlet puts it, and has to leave the room.

The play-within-the-play—like the play itself, and like the ghost, like a phantom limb, like language—is something unreal that has real effects.

Something's *viral* here: in laying the trap by putting on the play-within-the-play, Hamlet—who has been pretending to be mad—betrays himself as dangerously sane, which leads the King to kill him. Hamlet has caught himself in his own trap.

The power of this systematicization isn't confined to the play's "inside." Shakespeare's plays have been mounted as critiques of various heads-of-state who, "frighted by false fire," have betrayed themselves by their reactions.

We confront here not a play and a world but *a world made of interactions of real and unreal entities*—plays, personae, phantoms, performers, persons.

With all this in mind, let's return to phantom limb pain and the question of where the feeling takes place. For the brain, the body and the world are *imaginary*: they exist only as chemical/electrical signals it processes.

The brain is one step removed or withdrawn as if in a room watching the world
on a TV monitor rather than through a window. The world is imaginary;
it is always a *virtual* world, a *phantom* world. But this is only part of the story.

The brain *actually* depends on the body—if the heart stops beating,
the brain will die—but the body has come to depend on the brain,
and if the brain stops working, the body will die.

Magic lies in how the brain's phantom world participates in reality,
not by "representing" it but via the same simultaneous withdrawal
and heightened participation that are characteristic of all systems.

Like the phantom limb in which real feelings are located,
like the play (a tissue of lies, false fire that *really* scares us),
language *points to an absent presence*.

It is not something already there to which it refers,
but something *more than the sum of its parts*,
a real ghost that emerges from the way our brains
are wired to themselves, to others, and to the world.

If we could fully grasp the implications of this, the "hard
problem" of consciousness would cease to trouble us.

4   Biting Game

After a painful stomach infection at six months of age, a little girl
began to lie awake nights, crying. At nine months, she began to have fits
and to become anxious, jumping even at small sounds.

At a year old, she was having four or five fits a day and often cried inconsolably.
Her mother brought her to psychoanalyst D.W. Winnicot. Sitting on his knee,
the child made what Winnicot called "a furtive attempt to bite my knuckle."

In the next session, "she bit my knuckle three times
so severely that the skin was nearly torn" and then
"played at throwing spatulas on the floor, crying all the while."

In the next session, "she again bit my knuckle severely,
this time without showing any guilt feelings, and then
played the game of biting and throwing away spatulas;
while on my knee she became able to enjoy play.
After a while she began to finger her toes."

This led to "experimentation which absorbed her whole interest.
It looked as if she was discovering and proving over and over,
to her great satisfaction, that whereas spatulas can be mouthed,
thrown away and lost, toes cannot be pulled off."

Immediately thereafter, the mother reported she'd become "a different child":
she had no more fits, *not one*, and she was sleeping through the night.

When Winnicot saw her a year later, she still "had no symptom whatsoever."
She was "an entirely healthy, happy, intelligent and friendly child,
fond of play, and free from the common anxieties."

So what happened? Although Winnicot doesn't explain much, it seems
that simply by allowing himself to be bitten—presumably without pulling away
or scolding—he has mirrored what it's like to be an "indestructible object."

She has gotten the idea that pain need not be catastrophic and disintegrative.
She has confirmed this through play, and the realization has rewired her brain.

The rewiring affects proprioception (processing of internal data, such as pains),
sensory perception (the processing of external data, such as sudden noises),
and, engaging both, the child's ability to play and to enjoy and learn from play.

(Note that the possibility of play depends on the immediate stakes not being too high.
This lesson is delivered to the child by way of Winnicot's nonresponse to her biting—
or better, it's the answer he gave to the question he knew she was posing by her biting.)

And this is not just a one-way rewiring but also affects how the brain, in turn,
regulates the body (crying, sleeping, responding), the child's relationships
to others (and how they, in turn, respond to her)—in short, *everything*.

The magic is in how something so small and one step removed from the child—
his knuckle being bitten—has rewired *her brain*. It must be via some mechanism
(call it *mirror neurons* if you want), but this doesn't take away any of the magic.

Winnicot is able to perform the magic because, by understanding
what he and his patient are doing together as play—and play as reality-testing—
he's able to suppress the natural response to pull away when bitten.

In other words, his practice involves the continuous exercise of *empathy*
(admittedly a telegraphic way of talking about therapeutic practice):
by putting himself in his clients' shoes, he has rewired his own reactions.

This not only cognitively and affectively reframes
everything that goes on in the session: it has even
effectively overridden his own brain/hand wiring!

And notice that empathy is not so much an ethical imperative
as *itself a form of play*! This is a powerful reframing of ethics,
along the lines of the Romantic understanding of *imagination*.

Finally, although the therapist has been transformed by subtle and powerful knowledge
his patient has not yet fully developed, this does not go toward casting them in the roles
of authoritative doctor and passive patient, or powerful shaman and gullible supplicant.

His practice simply *allows him to play her game on her terms*.
And that, my friends, is where the magic comes from. At its best,
it's also a pretty good description of how science engages the world.

5   Dog Whisperer

A woman consults an animal trainer—a "dog whisperer"—
to see if he can help with her dog's extreme fear of heights,
which for years has meant that the dog refuses to climb stairs.

The trainer watches as she walks the dog on a leash,
and sure enough, when they approach some stairs,
the dog sits and won't let himself be dragged any further.

The trainer sees that the woman, anticipating the dog's reaction,
has made some advance adjustment in the tension of the leash,
and he suspects that this has *triggered* the dog's reaction.

He advises her to try again, avoiding any change in her handling of the leash as they approach the stairs. Sure enough, this time the dog follows her right up without missing a beat! After years of categorically refusing to climb stairs!

The magic here is not only, as with the Winnicot example, in making a long-standing and seemingly intractable problem disappear with what amounts to the wave of a wand.

As often in such cases, the dog's owner didn't understand how *she had become part of a system*. To her it seemed the dog's fear of heights came first—as it may well have. But her own anticipation repeatedly reinforced it, becoming wired into the system as a trigger in turn.

This is how a system grows: the response now comes *first*, and that which triggered it is now triggered *by it*.

6 Conclusion

Most of these examples involve a crucial and strategic *relaxation*. The mirror box and the Alexander Technique involve unclenching something that has been habitually clenched. The Winnicot example hinges on the therapist's ability to inhibit the tensing and reactivity of his hand, as the dog owner must refrain from tensing in anticipation. Even the nursing mother's laughter may be understood in this way.

The placebo experiment and the trying-to-get-pregnant example are exercises in the mindful *letting go* of both belief and disbelief.

This letting go explains why the placebo experiment was more effective than actual medication, which presumably was dispensed without any explicit suggestion that "you don't have to believe in it."

It is a very subtle difference, so please pay close attention: in everyday medical practice, all concerned are supposed to believe (without thinking much) that medicine works if they believe or not.

In the placebo experiment, all concerned are actively focused on something that will work through their minds but via the suspension of belief and disbelief, a ritual and an act of will that operates by a letting go: *an act of magic.*

CHAPTER 4

Future Perfect

1 The Gettysburg Address as a Magical Speech Act

The famous opposition between words and deeds in Lincoln's Gettysburg Address—
and the assertion that the deeds will live on but the words will be forgotten—
are both demonstrably *false*. The Address itself shows that the opposition is false,
that Lincoln well understood his words *as deeds* and knew they'd circulate widely.

So even at the time, to say "the world will little note ... what we say here"
might have been called *false modesty*, even if motivated by genuine humility
on the occasion of uttering words at a place where others had given their lives.

The perspective starts to shift when we focus on the *performativity* of the words
(that is, on *what they actually do in the world*) rather than on their truth or falsity.

If you think of the statement ("the world will little note") as attempting
to minimize itself in favor of the deeds, you'd still have to conclude
that its performativity fails along with its referential meaning.

It makes sense only when you see it as *counterperformative*:
a big part of *what the words do* in the world (making
themselves memorable) *is the exact opposite of what they say*.

Think of it as part of the magic, or the magic trick, of *misdirection*:
the speaker divides words and deeds, focusing the audience on deeds
so that the magic of *words as deeds* can better do their work.

This move derives from the foundational ruse (and magic) of language,
which seems to refer to things, but only *in order to do something*.
The referential dimension of language has always served the performative.

Lincoln's words refer to the past, to something that can only be echoed after the fact,
but in so referring, they do something *in the present*, where saying and hearing the words
directs us *to shape* and retroactively *to be shaped by* a very specific future.

The Address uses the religious and legal forms of *witnessing* and *taking an oath* ("we here highly resolve") to enact a performative ritual: by participating together in this event, *we here*, by simply receiving the words, participate in their enactment.

The words ground us in a past (in the materiality of a place and the deaths that have taken place here) and in a present (in social materiality, in our being assembled here together and in these words being uttered and heard by us all).

Most important, the words ground us in a particular future: a future in which "these dead shall not have died in vain." Because it imagines such a future as if it has already happened, the verb tense is called *future perfect*.

The words loop together a past, present, and future that shape one another. This catalyzing or co-shaping or temporal constellation is *what the words do*: they perform a magical or religious act that might be called a *consecration*.

This is not an act that inevitably sets a chain of events in motion (like striking a cue ball, the classic illustration of linear causality) but it is *an act of will that binds the present to a future result*.

The words do not *refer* to a looping together of past, present and future. They *perform it*, and insofar as the performance is successful, the words may come (retroactively, as it were) to be counted among its mediators.

The Gettysburg Address may be rare in its clarity, but the resolve mentioned in it, "that these dead shall not have died in vain"—that something noble should come even from carnage and unspeakable suffering—is a common one: *a will to meaning*.

2   Pool, Poetry, Prose, and Painting

Because the game of pool has long been used to illustrate linear causality (its popularity rose along with Enlightenment rationalism in the 18th century), it's a good place to show what can be gained by rethinking via nonlinearity.

The injunction to "be the target" (as per the 1948 book *Zen in the Art of Archery*) is one familiar way of aiming to replace linear with nonlinear causality.

It can be made simple: rather than seeing a shot proceeding step-by-step forward in time and space, the end point or result can retroactively organize the steps that lead up to it.

FUTURE PERFECT

In playing pool, you can't do well if you're looking at the cueball
(much less the tip of your cuestick) when you make your shot.
It's better if you look at the object ball, better still if you look ahead
to the pocket. The angles adjust backwards from the end point,
starting with the trajectory of the object ball into the pocket,
which determines where the cueball must strike the object ball,
where you must therefore aim the cueball, and finally, the position
of your hand on the cuestick and your stance. The whole sequence
is steered from the pocket back to your hands and feet.

And by the way, to emphasize *looking* is more than misleading:
vision needs to just help pass the message on to the hands and feet.
The eyes screw things up when they think they run the show.

In practice, you adjust back and forth as if the sequence were a set of linked angles.
Changing one changes the others (not to mention *spin*, adding physics to geometry).

ASIDE: Dr. Livingston's Magical Bank-Shot Visualizer

Just to show that this isn't purely theoretical, here's a tip for you.
People find bank shots difficult because the pocket is behind them.
They can't see to steer backwards from it, as described above.

When shooting bank shots, I came to realize that I feel
the table surface duplicated, as if it were connected to itself,
opened flat via a piano hinge along the rail I'm shooting into.
In order to hit it off the rail and into the corner pocket,
just aim the object ball at the phantom corner pocket.
Go ahead, try it. You're welcome!

(If you're new to pool, also note that, at the moment
the cueball strikes the object ball, they must be aligned,
as if they were adjacent beads strung on the phantom line
going from where they first touch to where they're going.)

Being attached to the end means allowing the means to fall into place behind it. This falling into place is why *relaxation* is necessary.

Here's another example: Think of how nonlinearity works when writing rhymed poetry. Do you think of what you want to say (*your desire*), then start squeezing it into the meter and rhyme (*the constraints*)?

As in pool when you shoot while looking at the cueball,
this will lead to a stilted performance. But if you're an old hand,
or in the groove, or both, the constraints and the desire *dance*.

Desire may be in the lead, but there is an *attunement*;
the constraints may sometimes even show you your desire.
This is related to the basic question of how you use language.

Do you get some preverbal sense of what you want to say,
then start selecting words, or do you have to rush headlong,
finding fully what you want to say only as you speak or write?

There is evolution and system-building in all of these activities.
Write enough in iambic meter and you start to think in iambs.

What, at the beginning, was the end (the goal, when you started
trying to squeeze your thoughts into iambs) is, at the end,
the beginning: part of the initial conditions of possibility.

When you're used to it, your syntax, as it approaches a rhyme, adapts itself
in advance, as an outfielder adjusts speed and direction continuously while
running to catch a fly ball, arriving at just the right time and place. Like magic.

It goes to show: brain, eyes, ears, and legs can solve nonlinear equations
when posed properly. In the case of poetry, both the possible pathway
and the destination change, adjusting continuously in relation to each other.

Rhymed poetry is a way of making such nonlinearity conspicuous:
we are enabled to *feel* the delicious knitting-together of a system.

And there is also crucial nonlinearity in writing (and reading) *fiction*.
Again, this is not subjective or theoretical or even very difficult.

I am simply referring to how, as a story evolves, the writer can go back
and write or rewrite what came before so the story knits together better
(no less than a reader constantly revises interpretations and expectations).

When William Godwin wrote *Caleb Williams* (1794), he first imagined
the situation of the latter part of the novel: a man is being pursued
by a relentless, oppressive power from which he cannot escape.

Godwin proceeded backwards, going on to posit
how such a situation could have come about:
a servant flees after finding his master's dark secret.

Then Godwin spun out the narrative forward again,
imagining possible outcomes. He went on to write
two opposed but both diabolically unhappy endings.

Was it cheating to work backwards, making it seem as if the situation unfolded
with a fateful yet unpredictable inevitability, giving it the feel of a real world?

It does fly in the face of novelistic *naturalism* (as developed in the 19th century),
whereby the author places characters in situations, then runs the algorithm forward.

As the story unfolds, we (like the author) are meant to discover—
as in a scientific experiment—how each character fares:
who evolves and triumphs, who goes down in flames.

Of course, the novelist already had in mind characters differentiated by feature-sets,
and a world that works by particular logics, rewarding particular features differently.

So naturalism's a set-up after all. The house always wins.
The cheating was in pretending the process could be linear
and exclusively forward-moving. Nonlinear composition
has the better claim of evolving as systems do, *naturally*,
by nonlinear knittings-together.

If you prefer a hard-science version, mathematicians Rajković and Milovanović
have developed a computational analysis by which the copy of a painting
can be distinguished from an original, even when the original artist made the copy.

This is possible because what the two mathematicians call the "causal architecture"
of an original work involves *self-organization*, a version of the looking-and-adjusting
backwards-and-forwards I have already described in pool, poetry, and prose.

Self-organization "denotes a spontaneous emergence of structures
and organized behavior without any external influence in systems
consisting of a large number of interconnected elements."

"Due to the feedback relations among constitutive components,
the dynamics of self-organizing systems is non-linear.
Self-organization indicates a spontaneous increase
in structural entanglement (complexity) of a system over time."

So, while copying an image is like a laborious paint-by-numbers,
creating an original work involves a feedback loop between an expected result
and the process of getting there, where both adjust continuously to each other.

As the irrepressibly nonlinear William Blake observed, "To Engrave after
another Painter is infinitely more laborious than to Engrave ones own Inventions."

For the viewer as for the artist, engaging the work involves a play with predictions,
a nonlinear process of toggling back and forth between expectation and observation.

As per the two mathematicians, the "interplay of predictive patterns
and unpredictable interruptions and the proportion of their occurrence
determines to a large degree the aesthetic experience and gratification."

Seeing, for viewer and artist, is not just a matter of adjusting to what's there,
but in the process, creating a new interpretive frame, a new relationship
between artist/viewer/work that shapes/reshapes all parties to the interaction.

3   Meteors, Messiahs, and Migraines

I announce that a meteor will strike the earth on a particular date
and throw the earth back into a new dark age. I start recruiting converts.
When the day comes and goes with no meteor, I issue an explanation.

Sorry, I miscalculated the date. Our little cult is thrown into crisis but recoups.
After several dates come and go, we decide the predicted strike was *metaphorical*.
In any case, because we've grown attached to the meanings and lives
we have built together, we find other grounds on which to continue.

There are other unlikely but possible paths to success.
A meteor, though not a gigantic one, *does* strike the earth—
on or reasonably close to the date I had predicted.

In fact, meteors at least four meters wide strike earth about once a year
(though most burn up in the atmosphere), and ten-meter-wide ones
cross our path about every ten years, so my prophecy isn't so unlikely.

Maybe a large meteor is reported to have just missed earth
(say, within 100,000 miles, half the distance to the moon).
Think of the validation! What a boon to our recruitment!

>ASIDE: On This Rock I Build My Church
>
>Astronomer William Hartmann has a theory
>about Paul's conversion on the road to Damascus.
>
>To Hartmann, Paul's story sounds a lot like accounts
>of the 2013 fireball meteor of Chelyabinsk, Russia.

>   Until his conversion, Paul had been a zealous anti-Christian,
>   but his beliefs must have had the capacity and readiness to shift.
>
>   Still, it's not hard to imagine that anything less dramatic
>   than a meteor might have failed to trigger his transformation.
>
>   So it seems that it took a meteor—a real entity sent from the heavens—
>   to deliver the good news: curious point of contact between Christianity
>   and animistic meteor-worship. Or *maybe it was just a migraine?*

The problem with meteors as the cornerstones of cults is that meteor strikes are outside our powers to effect. Better to proclaim some kind of messiah. Then we can argue whether he was the one, whether various miracles can be attributed to him, and so on.

We can even say that he knew he would be rejected in his own time
and that his adherents would be tested and would have to keep the faith
until the day of their vindication and redemption comes. Seamless!

This is as much a *validation* as it is a deflation of religion.

Just change the terms: in my vision, our destiny is to end slavery.
Unlike the meteorists, we have a chance of making this happen,
though it will be very difficult. We recruit converts to our cause.

Are you going to sit me down and explain that I have mistaken
the *performativity* of my vision—the extent to which believing
in it or willing it can help make it so—for its *referentiality*?

My answer is simple: *you're the one* who seems to be hung up
on separating "self-fulfilling" prophecies from "real" ones.

I could tell you that what you're so carefully calling
"the extent to which willing it can help make it so"
is what we call "how God or Destiny works through us."

I could tell you that, if the self-fulfilling prophecy works,
it *will have been real*. But my concern is much more basic:
are you with us or against us? Well?

ASIDE: Falling Off a Log

Here is a much humbler example of willpower and performativity.

As long as I kept smoking, I noticed that every single thing I said
about my habit, positive or negative, was, basically, *bullshit*.

As long as I was still smoking, everything I said served the habit,
whether by rationalizing or enabling me to beat myself up about it.

When I finally quit, I liked to say *it was easy*, like falling
off a log (though I had to relearn how to do other things,
such as staying calm, writing, getting through the day).

The exercise of will seemed absolute to me. I compared it
to cutting off your foot: it's painful and you have to hobble,
but there's no sewing it back on.

So the end point was *where I started*:
everything had to fall in place behind it,
in its wake, sometimes turbulently.

When I said it was easy, people thought I was being sarcastic, or in denial,
or defiant, or something—but it didn't seem to me that this was the case.

As long as I didn't smoke, anything I said about it was *true*—
or, to put it more accurately, *performatively happy* or *effective*.

I should publish a self-help book: *Dr. Livingston's Quit Smoking Forever*.
Chapter One would be *The Easiest Thing You Ever Did*. Chapter Two:
The Difference between Willing and Wanting. Three: *How To Find
the Portal between the Smoking and Non-Smoking Universes*. Four:
Angels That Guard the Portal. Five: *What To Do When Haunted
by the Proximity of Other Universes in Which You Still Smoke*.

I had tried readily available narratives and metaphors.
They didn't work. I don't *believe* in an angel-guarded portal,
but something about it seems right—and *it worked*.

> Curiously effective metaphors warrant study
> not to demystify and deflate their metaphoricity
> but to learn to use *their curious magical effectivity.*

4   Magical Militarism

As we saw, the grounding performed by the Gettysburg Address
is the binding back of a future to a present. It is important not to say *the* future,
as if there were only one, or to say *the* present. There are just as many presents.

In some presents, we are engaged in an epochal transformation;
in others, the speaker is a clumsy rhetorician; the audience dupes.

The *Chicago Times* asserted that "the cheek of every American must tingle with shame"
upon reading Lincoln's "silly, flat and dishwatery utterances." The *Times* of London
found that "the ceremony was rendered ludicrous" by Lincoln's "luckless sallies."

Because meaning is continually under construction, there is always time
for the meaning of the moment to change, convulsively or gradually, *in retrospect.*

Most are now inclined to say that Lincoln succeeded in construing the Civil War
as about the recognition of a common humanity, but it is also easy to imagine
iterations of this nation—even present ones—that mock this construction.

The nation that elected a black president is still characterized by ongoing acts
of state and individual terrorism against black people by white supremacists—
on the streets, in churches, by police. A full range of kinds of slavery still exists.

Arguably, in the fullness of time, we could also be revealed to have been dupes
in the fullest sense: as a gigantic meteor speeds unstoppably toward the planet,
where's your precious meaningfulness? *Now* will you join our meteor cult?

You could say that meaning is trumped by time, which always gets the last laugh.
Therefore, while you're at it, because you'll eventually die, why get out of bed?
If this sounds right, consider making an appointment with a psychopharmacologist.

A past, present, and future can only be knit together *for a finite amount of time.*
That such a knitting eventually unravels is not a failure of meaning but an inevitability
for all living things. Sustainability is finite; the idea is to prolong, not to immortalize.

The Gettysburg Address seems to occupy the tragic-heroic battleground between
prolonging and immortalizing (between "can long endure" and "shall not perish").

So finally, the resolve "that these dead shall not have died in vain"
is not only part of some universal human will to meaning.
It is common to warmongers in all times and places.

The Address works by deploying the core patriarchal tropes of *male birth*
(established in the phrases *fathers brought forth* and *conceived in liberty*)
and of political entities as *alive* (as in *nation might live* and *shall not perish*).

These tropes tend to go along with notions of renewal—and ultimately
immortality—to be sought through warfare, violence and death.
Regeneration through Violence is a long-standing national myth.

The myth encodes a masculinity that is fundamentally tragic,
in which we fall, struggling for an ideal that forever recedes before us.
This tragic mode, sometimes also *hysterical*, seems to trump everything.

Dying in battle trumps living for others. Men giving birth
to nations trumps women giving birth to actual children.

Even as I recognize that Lincoln's oratorical magic participates in this myth,
a big part of me recoils from the implications of this analysis. I hear a voice
(incredulously): *You're saying there was a nonviolent alternative to the Civil War?*

While you're at it, how about a pacifist movement that could have defeated the Nazis?
And, from a safe distance of course, let's tell people everywhere surrounded by others
who believe they have no right to exist that they should turn the other cheek.

I've resisted assigning these remarks to a heckler from the sidelines.
They come from part of me that recoils from what seems dangerously naive.

That part wants me to remind you that what it calls "I" *earned its defensiveness*
by being raised under the constant and active pressure of psychological negation
(however this may pale next to other kinds of abuse and physical violence).

So, then, the task is to resist aggrandizing or negating this part in turn,
and to recognize it—with compassion rather than forgiveness, I want to say—
for what it is (*mark of weakness, mark of woe*) and not be pulled into its orbit.

It is the still-toxic psychological soil of patriarchy
where the seeds of white supremacism, jihadism,
and other more everyday forms of violence grow.

Could a nonviolent movement have succeeded, maybe winning hearts and minds
of white southerners better than military defeat could have? But even if so,
wouldn't it have condemned more generations of African Americans to slavery?

Of course I can't say, not just because I am not a historian, but because
the inevitability of violence and war has been encoded so deeply, broadly,
intensely, continously—if counterfactually, ideologically, and magically.

The question of whether there could be a *science of ideology*
so thoroughgoing as to defuse the myth of inevitable violence
is not opposed to the question of whether there could be
a magical nonviolence compelling enough to do the same.

Read the previous sentence as many times as you have to.
I tried to say it more simply, but if you want to juggle,
at some point, all the balls have to get into the air.

Could there have been a nonviolent alternative to the Civil War?
If we can't yet engage such a question, let's take a step back and ask,
instead, in what future will another past turn out to have been possible?

From what standpoint could we say, with Walter Benjamin, "As flowers
turn toward the sun, by dint of a secret heliotropism the past strives
to turn toward that sun which is rising in the sky of history"?

In which future will we here, in this very moment, right now
(the two very different moments you and I each call this moment),
have been engaged in an epochal transformation?

5   Four Asides

    ASIDE: Dr. Livingston's Time Travel 101

    History is generally not taught in the mode of *what if*.
    But what if teaching history worked to find conjunctures
    in which things could have gone some other way?

These might not be discrete moments but constellations of factors
operating together that could be leveraged at what might be called
acupressure points. Call it Speculative Historical Acupressure.

This is my method for training speculative historians of nonviolence.

Your assignment is to time-travel, with a team of your own choosing,
back to whatever moment you identify as the optimal intervention point,
and end slavery in no more than a generation *but without the Civil War*.

We will have prepared for this difficult assignment by other easier ones,
such as time travel to recover lost works of art, prevent the conception
of Hitler or Dick Cheney or other assorted war criminals, and so on.

What would it mean to go back to prevent Christ's crucifixion?
If Christ *willingly* made the sacrifice, wouldn't preventing it
be the most radically anti-Christian act imaginable?
Would it even be possible, in a theological or historical sense?

I told you the assignments would get progressively more difficult.

Who would you consult before going? Who would you bring with you?
What would you do? What political, ideological, religious maneuvers
might stand the best chance of leveraging the situation? Would you write
broadsides or books, find a way to whisper stratagems to key players?
Would you reveal that you were from the future? Would you go back
to the drafting of the Constitution? To before the slave trade began?

We'll select which proposals the Time Travel Bureau would sponsor,
cherry-pick the best ideas, and collaborate on a second iteration.
Once we beta-test, we can turn it into a national online competition:

gamified, crowd-sourced, Speculative Historical Acupressure.
Of course, it is theoretical and mostly illegitimate as history,
but would it work as a kind of simulation training for activists?

What would the proposals look like after twenty years of classes?
What other spin-offs in other disciplines can you imagine? Could we work
this way with scientists on climate change and even evolution?

How would the concept itself evolve? What if nobody
takes up Dr. Livingston's proposal in the first place?
How would the world be different? (And by the way,
next semester, in Time Travel 103, we'll go to *the future*.)

ASIDE: A Somewhat Rationalizing Account of Mystical Nonviolence

Percy Shelley's 1819 "Mask of Anarchy" is one of the most radical poems
ever written. It helped Thoreau define the practice of Civil Disobedience,
which in turn influenced Gandhi, who quoted the poem often. In the US, it
was used in labor organizing and in the civil rights movement.
Chinese students chanted it during the Tiananmen protests in 1989.

Not bad for a poem that Shelley couldn't even get published
by his best friend, who happened to be a newspaper editor!

At the poem's turning point, a mysterious and ethereal Shape
with various mythic attributes emerges from the sky
and steps lightly over a dispirited crowd of protesters.

The Shape moving over them organizes and galvanizes them
with the principles of nonviolent resistance, enabling them,
somehow, for the first time, to stand up to their oppressors.

Never has so much ridden on a *MacGuffin*, on a *deus ex machina
so airy-fairy*. Shelley mobilizes a flurry of hyperpoetic similes
to describe how the Shape moves above and through the crowd.

> As flowers beneath May's footstep waken,
> As stars from Night's loose hair are shaken,
> As waves arise when loud winds call,
> Thoughts sprung where'er that step did fall.

To understand the operation of a Romantic symbol like this Shape,
you have to set aside the idea that it might be a metaphor or a simile
for something (for the "Spirit of Revolution," or whatever).

The Shape, when the poem circulates and is read,
and if and when it inspires its readers,
will turn out in fact to have been *the poem itself*.

The poem will turn out to have been the phantom linking
Shelley's revolutionary desire with that of his readers as it steps—
lightly—across the real-space networks of its distribution as a text.

The poem steps across the synaptic and social networks
of its individual readers, where indeed "thoughts spring up"
(as Shelley said) wherever it circulates and is read.

Shelley had intended for the poem to be published in a newspaper,
where it would be read one day and then, presumably, discarded.

Indeed, when in the poem the protesters look around
for whatever catalyst has galvanized them, whatever
it was has disappeared: "all was empty air."

This ephemerality makes it clear how the Shape operates
by a butterfly-effect triggering that can never be found
by tracing back to causes. This is the magic of *textuality*.

Shelley's account of the Shape's iridescent wings and soft step
as it passes by overhead makes it quite butterflylike; more evidence
for how long *Butterfly Effect* had been in the air before being named.

Though Shelley put it grandiosely, all this tends to bear out his claim
that poets can be "unacknowledged legislators of the World,"
"mirrors of the gigantic shadows futurity casts upon the present."

ASIDE: A Somewhat Mysticizing Account of Rationalist Nonviolence

Political scientist Erica Chenoweth found that, over the past 100 years—
and across the world—nonviolent campaigns to overthrow governments
(vs. violent ones) succeed and bring democratic change twice as often;

that every movement that got sustained participation
from at least 3.5% of the population succeeded;
and that *every single one* of these was nonviolent.

As Chenoweth explains, she began by assuming that violence and war
were, regrettably, necessary. Her story is a classic scholarly narrative
in which science triumphs by overcoming the "bad magic" of ideology.

If you prefer, we could say instead that she had a *conversion experience*.
Unlike the one triggered by the meteor that may have shocked-and-awed
Paul on the road to Damascus, this one was more subtle and sustained.

It began, she says, when an activist challenged her
to research the efficacy of nonviolent movements.

But what was it that spurred the activist to challenge her?
What spurred her to follow through? And what, after all,
was it that held the nonviolent movements together?

In all of these cases, was it, at bottom, political rationality,
or something more like religious conviction or magic?

ASIDE: Four Scholars

A botanist works for years in obscurity, shunned or ignored
by most of her colleagues. She believes she has discovered
a slowly unfolding, thoroughgoing ecocatastrophe in process.

This crisis is unrelated to climate change, but proceeds instead
from microscopic causes via a domino effect so subtle and circuitous
it's easy for her detractors to dismiss her as a paranoid doomsayer.

All the while, tormented by doubts, she keeps
believing steadily in the import of her work.
In her later years she sinks into dementia.

When she is visited by one of her few true believers,
an old graduate student, she no longer remembers who she is,
much less anything of her work, and she dies shortly thereafter.

Decades later, one of her few published texts, languishing unread
in some now-long-defunct journal, is discovered by a biologist
who realizes that the world-we-know might still be saved by it.

This second scholar writes up his findings and is about to deliver them
to a high-placed friend who has the power to act on the discovery,
but on his way he crashes his car and is killed.

After many more decades, the world is sinking into chaos.
The ecological collapse the first scholar predicted is in full swing,
though the microscopic triggers that set it in motion remain unknown.

Eventually a third scholar finds (in an obscure archive)
the second scholar's work on the first scholar and realizes
the course of world history might have been transformed by it.

Fired up by the philosophical implications of this revelation—
though it is too late now for it to be put to any practical use—
she starts what becomes a new kind of religion based on it.

The religion begins to spread, unifying isolated camps of survivors
under the banner of noble new ideals, when the earth is struck
by gamma-ray bursts from a nearby supernova, wiping out all life.

Thousands of years later, aliens on a scouting mission land
on the scorched planet that once was Earth. When they go home,
the third scholar's religious tracts are delivered to an alien scholar.

This scholar begins to decipher the tracts. Late one night, sitting
under a huge arched window, drinking coffee (yes, the aliens
have something like coffee that they revere as a sacred beverage),

immersed in deciphering and lit by the tremulous light
of two of his planet's five moons, this scholar is struck
by a sense of cosmic melancholy known as *b'sha-ah*.

He watches as a trifenkulenx climbs up a table leg
and up the side of his blue coffee cup, pauses at the lip,
its three antennae quivering, then climbs back down
the cup, down the table leg, and off into the shadows.

To come to the brink and to turn away from it!
Not to fall into the cup and drown, flailing. Not to descend
back into the shadows without ever having visited the edge!

The scholar has no idea if the bug recognized danger
while on a scouting mission or simply strayed there
and wandered off on his way again just as aimlessly.

Without being able to say what revelation has been delivered to him
by what he has witnessed, he is struck by humility and gratitude
for being a creature of a world sustained by magic and miracle.

The little bug was the catalyst for his revelation,
and had he not happened to turn and see it,
who knows if his world would have been rocked?

Though it was the trifenkulenx that delivered the revelation,
it could never have happened without the work of the three scholars
who preceded him, or the melancholy fate of their entire planet.

That something so tiny and so random should conspire with something
so epic, the scatological with the eschatological! But what happens now?

Will the fourth scholar find a way to articulate his revelation?
And if so, will it be in the form of what we would call science
or philosophy or religion, and how will it, this time, be received?

Or will the moment of sublime clarity simply pass, as such moments do?

All our work might come to naught. It might be derailed
by the most random, the most petty of circumstances—
or hang on by the slenderest thread and thrive against all odds.

And what's more, one of these doesn't preclude the other.
Unbeknownst to us, it may, in the fullness of time,
save the world or worlds yet undiscovered.

And yet, in the fullness of time, all must come to naught anyway, yes?

CHAPTER 5

What is Religion?

1   A Fuzzy Set

Belief in god is not the focus of all religions. Buddhism is a good case in point.
Even *belief* in general is not the issue for millions of people worldwide
who identify with a religion ethically or ethnically but do not hold its beliefs.

But you can also find lots of people who profess
the beliefs but don't otherwise practice the religion.

So ... if religion doesn't focus on a god or gods
or even on specific beliefs or practices, *what is it?*

The easiest real-world answer begins by recognizing that religion,
like most concepts, is a *fuzzy set* based on *family resemblances*.

In other words, everything that counts as a religion
will possess some but not necessarily all of a set of key features.
As a first approximation, then, let's try to identify such features.

Religion provides a framework (some *higher plane*)
that offers to give our lives and worlds *meaning*.

It identifies entities and forces and laws that aren't reducible
to the immediately obvious beings and actions before us.
It features sets of practices (and, often, texts) that serve
to keep us mindful of this framework and these entities.

It seeks to engage these entities directly—often to intercede with them
and get them to intercede on our behalf. In doing all these other things,
it binds its adherents to each other as well as to a particular kind of world.

Does this sound about right, for a start? But notice
that, while trying to generalize the features of religion,
I've also managed to give a pretty good account of *science*.

In science, the *non-obvious entities* are things like electromagnetism and quarks. Such entities figure into an account of the totalizing framework called *nature*. The practices that address and deploy them are *scientific method* and *technology*.

The account also works for politics, whose entities are nations, parties, and so on. These entities are not obscure in the same way, but their abstractness and expression in *laws* give them distinct family resemblances to religion and science.

To start defining *magic,* we could point to the same positing of non-obvious entities and forces and the sets of practices and interventions that go with them.

The term *occult* (*hidden*) emphasizes their non-obviousness, but this feature is shared with religion and science, which have their own *arcana* and *adepts*.

2   Magic, Science, and Religion Coevolve

From a historical perspective, resemblances among science, magic and religion should not be too surprising: the three coevolved.

In 16th- and 17th-century Europe, some alchemists became chemists, and some strands of *magic* became *natural magic* and *natural philosophy*, later known as *science*. What look now like amalgamations of occult and scientific societies, such as the *Academia Secretorum Naturae* and Galileo's *Accademia dei Lincei,* arose in Italy and elsewhere. In Britain, secret societies such as the *Invisible College* paved the way to groups like the *Royal Society of London for Improving Natural Knowledge*, founded in 1663; alchemist and chemist Robert Boyle was a member of both.

The Church—rightly and wrongly—often understood science and magic as part of the same heretical enterprise.

Historians of science have recognized that science also began to take shape *as a form of politics*, a way for the rising middle class to claim absolute authority of its own, in the name of its own invented entities, Nature and Reason.

Thus science did its best to dodge direct confrontation with the entrenched and absolute authority of *church* and state, and with their powerful imaginary friends, *God* and *King*.

The history of science is also interwoven with stagecraft
and magic-as-entertainment. Well into the 19th century,
those we now call *scientists* put on for the public
showy demonstrations of newfound phenomena.

And magic acts also incorporated new scientific technologies.
Phantasmagoria and magic lantern shows gave way to Fantascopes,
Kinetoscopes, Cinematographs—and eventually, cinema.

Objects of magical knowledge also became *scientized*. Renaissance mages
(as per anthropologist Stanley Jeyaraja Tambiah's insightful account)
"turned to number symbolism and mathematics as the key to operations."

"Playing with number symbolism paved the way for mathematics proper,"
and "the subsequent trajectory of theoretical and applied sciences
has vindicated mathematics as one of the master keys by which
the forces of nature can be manipulated and harnessed."

The study of magnetism, gravity, electricity, light, and heat
all advanced by postulating some kind of occult fluid, or *ether*.
This term continued to be used both by spiritualists *and* scientists
into the 20th century, when it was displaced by the term *field*.

3 Restoring the Chaos

To make sense of these entangled histories, the usual account
relies on the pervasive metaphor of a tree. Once upon a time
(so the story goes) a single worldview prevailed.

Evolution proceeded in the form of branching and differentiation:
science, religion, politics, magic, and art emerged from the mix,
evolving in their own directions—the modernist story of modernity.

As with all dominant metaphors, this one enables us to think through certain aspects
of the story at the expense of others it makes unintelligible. Wherever a tree metaphor
has been dominant, the countermetaphor of a *rhizome* gives some rethinking room.

Even when there are sharp differentiations among magic, religion, and science,
there's a sprawling, lively network of subterranean connections and resemblances.

The history of sexuality provides a curiously relevant and schematic example.
The terms *homosexuality* and *heterosexuality* were coined in the 19th century.

There is no doubt that this new binary notion of sexual identity
made unintelligible much of the detail of sexual practice,
with broad-ranging impact both on practices and social identities.

The range of practices *continued*—even as their meanings and linkages
to emotional and social complexes shifted. The new categories didn't fit everyone.
Some were allowed to wear them loosely. Others were straightjacketed by them.

In any case, sexualities—even the most vanilla ones—continue to be constituted
by slipperier and more sprawling linkages "than is dreamt of in your philosophy."
Look more closely, and the two categories break down into shifting constellations.

For starters (as in Eve Sedgwick's account), for some straight and some gay people,
sexuality is connected with sustained emotional intimacy, economic inter-dependence,
and/or child-rearing. For some it isn't. Depending on whether these go together for you
or not, you may share more with people *across* the homo/hetero divide—with whom
you may be watching your child at the playground, or looking for love on the internet.

As my friend Jack Halberstam suggests, the liberatory project
is to *restore the chaos,* and in the process, to create something
more like sexual, social, and political democracy.

As with sexual identity, we have been repeatedly offered "two-state solutions"
to the science-and-religion question. Evolutionary scientist Stephen Jay Gould
proposed that religion and science be regarded as "non-overlapping magisteria."

A *magisterium* is a realm ruled by one kind of authority.
Gould assigned religion *the realm of ultimate meaning*
and science *the empirical domain of facts.*

This resembles familiar dichotomies of supernatural/natural and faith/reason.
Which is unfortunate. As with sexuality, binary division offers clarity at a cost
of *making vital complexities unintelligible, getting stuck, and creating backlash.*

It would open things up a bit to think of this binary as part of a larger *plurality*
(which, I think, was where Gould was headed), but as those who study systems know,
singularity and plurality—and autonomy and dependence—are never absolute.

Autonomy always occurs with interdependence. I can walk around pleased
with my relative autonomy, singularity, and self-reliance because I'm a creature
dependent on other (plural) creatures, inside a kind of larger (singular) creature.

That creature is the living layer that covers our otherwise rocky planet,
sometimes known as *Gaia*. Systems emerge by increasing *both* the autonomy
and the interdependence of their components and relationships to other systems.

The non-overlapping of religion and science is *a ploy*, something you'd propose
when my magisterium just attacked your magisterium—a Trojan Horse.
Did Gould really think religion would retreat to the realm of ultimate meaning?

Or that science would be content with "just the facts, ma'am"?
In fact, science and religion encroach on each other's domains,
dreaming and scheming of various ways to subordinate the other.

This is not surprising, because there is something disingenuous about secularism
and the tolerance it preaches. Whatever I may say about respect for your beliefs,
if I believe in God, then I must believe that He is also your and everybody's God.

He must be your God even if you (defiantly) refuse to believe in Him.
Otherwise he wouldn't be God at all but merely *middle management*,
or the aspiring, petty warlord of some gerrymandered fiefdom.

Just so, if I'm a thoroughgoing rationalist, I'm likely to believe
that my rationality is *The* Rationality (and, normally, that
thou shalt have no alternative rationalities before me).

We confront a seamless, uninterruptible kind of *closure*, a closed-mindedness
often associated with religious people. But it is akin to—perhaps the same as—
the equally unassailable arrogance and condescension of mono-rationalists.

In fact, all systems (including belief systems) necessarily have some kind of closure,
though as explained in Chapter 2, the closure is, significantly, only *almost* total.

Recognizing your system as one of many—known as *pluralism*
or *relativism*—doesn't necessarily make you humble for long.
In fact, it turns out to be a very good way to feel *more* arrogant.

"I see that there are multiple frameworks and that mine is just one of them—
unlike you, stuck in your own framework! My pluralism makes me superior."

This is a big attraction of secularism. It enables us to identify *others*
as closed-minded, stuck, and narcissistic—like those primitive peoples
who are said to have no other name for themselves but "the people."

This must be one of the triggers for the attacks on relativism:
not that it's too thoroughgoing, but that it retains and intensifies
the arrogance it condemns (or righteously tolerates) in others.

I may say that "we are all the products of our own particular frameworks,"
but *in the act of saying it*, I seem to situate myself above the frameworks.
This is *counterperformativity*: words *doing* the opposite of what they *say*.

What they perform is known as *othering*—attributing to others what
is disavowed in the self. Disavowal is expert at covering its tracks.
I acknowledge *your right to your delusions*—I mean, *your beliefs*.

And I know that even in my own highly evolved cosmopolitanism,
I can sometimes be intolerant, and knowing that is what makes me
so smug—I mean, *so humble*. Sorry, what were you saying?

Fortunately, there is another way.

4   Rain Dance

Early monotheism—Judaism and Christianity—saw gods of competing religions
as *actually existing* and their rituals and magic as *genuinely powerful but evil*,
hence their commitment to an ongoing, high-stakes battle to defeat paganism.

In the mono-rationalist dogma of modernity, magic and primitive religion
are no longer characterized as *bad religion* (*bad* meaning *evil*) but caricatured
as *bad science* (*bad* = *mistaken*): misguided attempts to control the environment.

One of the most common examples is the rain dance.
How could people believe in the efficacy of such a ritual?
We humans are narcissists with control fantasies (the story goes)
so good at seeing patterns that we see them where they are not.

We are always ready to believe that rituals such as rain dances
are more effective than chance could explain because our "rage
for order" allows us easily to discount negative results.

When the rituals don't work, we can say they were misperformed,
or that someone or something else must be magically thwarting us.

But even evolutionary theorists have proposed that religion and ritual *work*,
that they can be *adaptive* in a strictly objective, scientific, evolutionary sense.

Religion works to produce *group cohesion*, we are told, by focusing
the group on tasks at hand and uniting it against other groups.

It works via the *performativity* of the rituals: the fact that,
in being performed, they actually *do* produce group unity.

This is related to their *self-referentiality*: the fact that, whatever else they represent
(e.g., *rain gods*), their main import is to characterize the group that performs them.

The rituals produce in the process a linked collective and
individual *self*. But even so—by this account, anyway—
the job of group cohesion is performed via a fundamental *mistake*.

Performativity or self-reference is mistaken for *referentiality*—
the reference to something external and supposedly really existing
in the world. Self-fulfilling prophecy is mistaken for real prophecy.

The name for the mistake of seeing referentiality where performativity
is in operation is *God*—and more broadly, *gods, magic, ritual,* and so on.

But as Wittgenstein asked, if the purpose of the rain dance
is to bring rain, and its practitioners believe it to be effective,
why don't they perform it in the dry season as well? *Well?*

A related question troubles anthropologists: How can shamans practice trickery
and still believe in their magic? We need to rethink. As a thought experiment,
put the rain dance into evolutionary context, and a different picture emerges.

The rise of agriculture was part of a linked constellation of developments,
starting with an increase in social group size and in the size of our brains
and, crucially in tandem with both of these, the emergence of language.

It doesn't make sense to ask which came first or which enabled the other.
They emerged as a system—each catalyzed and was catalyzed by the others,

along with other especially human adaptations. As scientists are realizing, there is a large constellation of such adaptations, including laughter, music, fire and cooking. In addition to other utilities, these served group cohesion, enabling people better to organize together and to attend to each other.

To become a more settled and more agriculture-dependent people, as we clearly did, you must do more than recognize the far-reaching meaning of the rain for your life.

You must tie yourself and your fate to the rain, submit to its agency and vagaries, attend ever more carefully—as if your life depended on it—to its habits and moods. This tying-together and giving-of-agency is what the rituals perform.

Tying together means becoming interdependent: humans become bigger players in the water cycle, in what the rain *does*. We become more important to the rain, but if we are all gods—as the saying goes—the rain is still *a bigger god*.

> ASIDE: From Flow
>
> Thermodynamicist Adrian Bejan proposed what he calls the *constructal law* as a fundamental natural principle.
>
> Put simply, what drives water—its dynamical prime directive (as for heat, electricity, and so on)—is *to flow and dissipate better*.
>
> In the elaborations of this process, water generates a host of living entities (themselves largely water) that act as new ways for it to flow and disperse.
>
> In this view, it is no metaphor to say that *water created us* in its relentless drive for self-realization and self-dissipation.

Let's say that *the magic* is in the emergence of this system, and that, as the system becomes more and more established, the rain—though it will always be the bigger god— really does become more and more dependent *on us*.

*W*elcome to the Anthropocene.

So it seems that those wily primitives were right after all, in every sense: their rituals were effective in binding them not just to each other but to the earth and sky in new ways.

This is a simple fact: they got *results*, transforming in the process
not only themselves as a people but also the actual earth and sky
(just as bacteria and plants had done a few billion years earlier).

Rather than simply rebounding on a self, self-reference and performativity
knit a system and environment together, operating at the interfaces of the two.

(As I wrote that last sentence, my fingers typed the word *togetherer*: good way
to think about the *intensification* that constitutes system/environment bonds.)

The primitives were most far-reachingly right—right in a way that testifies
compellingly to the efficacy of religion and magic—*in the fullness of time*.

They succeeded not only in producing particular crops in particular years,
but in setting us on a developmental pathway to a particular set of futures
for humankind and the planet. Cult, cultivation, agriculture, culture.

We became important not by simply imposing our will on the rain
(we haven't ever been able to do that) but by binding our fate to it—
something we must continue to recognize, or forget at our peril.

This is the forgetting that characterizes modernity,
a reversed instance of seeing only referentiality
where performativity is in operation.

In Latour's terms, the mistake is seeing "Nature One"—facts as if they existed *out there*,
independent of us—but not "Nature Two," which emerges from facts *found* and *invented*,
from our highly selective interactions and negotiations with nature and each other,
and from selective embodiment of our *curated facts* in technologies and other practices.

The symmetry shows how *the modern account of the premodern mistake
is simply the unrecognized mirror image of our own disavowed mistake.*

While studying the world—the truths supposed to reside there—
we have all along been altering ourselves and the world,
by how we have collectively wired together our brains.

If the primitives screwed up the science, we've been screwing up the magic.

What's the next phase, then? Some dialectical synthesis—
call it Magic Science Religion—that rectifies and transcends?
Or to be less Pollyannaish, can we make *a new mistake*?

Sure! But keep in mind that *Minerva's owl flies at twilight*,
meaning that only when a historical phase is ending
can we start to get perspective unavailable from inside.

If that's the case, we could only understand what new mistake we are making
if we were already in the process of displacing it, *making even newer mistakes*
with which we have yet to come to terms. Sounds about right, doesn't it?

If the self-aware synthesis implied by the term Magic Science Religion
is not quite possible, then say instead that it is the *grand new mistake*
I'm suggesting that we make and which we can *sustainably fail to realize*.

Like all failures and mistakes, this one has a directionality but not a destination.
It would take us individually and collectively down some paths and not others.

Is it just coy to say one could recognize something as a mistake
and still advocate it? Wouldn't it be like telling someone
you were giving them a placebo and still expecting it to work?

Wouldn't it be like maintaining that, rather than differentiating as they evolve,
creatures could merge; that sameness and difference could somehow overlap;
that $A = A$ and $A \neq A$ could both be true? If you read the last two chapters,
you may recall *that's just what I've been saying*—or if not, can you account
for what happens when you operate in the mode of sustained questioning?

5   Providence, via Vico

Giambattista Vico's 1725 *New Science* is the granddaddy of social sciences,
as they came to be called (anthropology, sociology, theology, and so on):
an early systematic study of how we humans make our worlds.

Because its fundamental principle of dialectical worldmaking
is so powerful, readers—myself included—are often bewildered
at how all that Vico spins from the principle seems wildly wrong.

This bafflement is where I start: I know my response ("wildly wrong")
is presumptuous—wildly wrong itself—but the question of what
makes reading *New Science* so vexatious *gives me a way into it*.

Then again, we often encounter theories in which the particulars
seem sound, even when the fundamental principles are all wrong
or just seem to be wrong in their counterintuitive strangeness.

The classic example is how premodern astronomers,
using a (wrong) earth-centered model of the universe,
were still able to make accurate planetary charts.

Though their predictions were bedeviled by inexplicable glitches,
the model could usually be tweaked to accommodate them (famously
by the proposition of smaller orbits-upon-orbits known as *epicycles*).

It goes deeper: the principle of *reductionism*, guiding
much of the most productive modern scientific work,
doesn't stand up even to basic philosophical scrutiny.

For example, reductionism tends to present itself as deductive
("seeing is believing") while denying the way its investments
in particular paradigms distort its world ("believing is seeing").

In fact, models and paradigms—metaphors able to achieve a degree of universality—
are what Vico defines as *poetic*s, which he regards as the source of all knowledges.

Even now, paradigms are usually understood as *heuristic*—
as ways of proceeding or of productively grouping things,
but not as anything like truths or facts in themselves.

Atoms aren't tiny solar systems, but one can get *pretty far*
by interacting with them on the basis of this assumption.

One paradigm is *as real* a way of proceeding as another:
it can't help but shape *selective making and interacting*,
even if what one makes, sooner or later, is a mess of things.

Wait awhile and all paradigms reveal their counterproductivities:
welcome to the Anthropocene. Such worldmaking is Vico's subject.

Especially here in the Anthropocene, one may find it hard to understand
how Vico could have explicitly and entirely set aside *nature*, as he did,
in favor of studying human social and cultural self-making.

A case could be made (in hindsight, a fanciful one) that if knowledge
had proceeded in a Viconian direction, we wouldn't be in this mess!

But in marking off from nature the social sciences, Vico
joined the nature/culture distinction central to modernity.

This demarcation was the other side of the constitutive mistake
made by the modern sciences of nature, which tend to ignore
how knowledges are shaped by society and history.

But we co-make not only ourselves as we produce particular knowledges
(socially and individually), but also *a particular kind of world* in the process—
a collaborative and conflictual process in which we are not the only players.

This is to expand Vico's core principle, starting with the recognition
that language, thought, and knowledge coevolve in dialectical loops
with our bodily, social, and historical interactions and environments.

So why does what Vico spins from this powerful principle seem so fanciful—
the stories of the giants who once inhabited the earth, the heavily schematized
stages of universal history, the division of everything into threes, and so on?

> ASIDE: Systematizers Systematized
>
> The other thinker who obsessively divided everything into threes
> was Charles Sanders Peirce, the philosopher who invented *semiotics*.
>
> As with Vico, Peirce's core principles have been transformative,
> while his many obsessive systematizations based on them
> have gathered dust (and hindered engagement with his work).
>
> My old teacher David Halliburton used to categorize thinkers
> according to *how many things they have* in their universes.

Monists have only one thing (usually spirit *or* matter)
that alone assumes all forms. They tend to be mystics.
Dualists have two things (like spirit *and* matter).

Dualists tend to be *stuck,* as Halliburton put it,
but he may have characterized them so negatively
because (as he asserted) he was himself a Three.

The Threes are *historicists*: his exemplars, Hegel and Marx, start
with dialectical process, (1) thesis, (2) antithesis and (3) synthesis.

The dynamism of Threes is of a radically different order
than dualist stuckness. Vico and Peirce built philosophies
on adaptive evolution and dialectical transformation.

The *Fours* tend to be mystics again: my own exemplar is Blake,
who, like Vico, found the source of everything in poetics,
but unlike him, or any 1 or 2 or 3, starts with a *fourfold* God.

Blake was also a kind of mad systematizer and schematizer.
His epic poems have led scholars also so inclined (I'm not)
to produce charts and diagrams that purport to decode them.

I *am* inclined to say that oneness, twoness, threeness, and fourness
are *translucent,* so you can see all the others inside each of them.
But then, so my fourness (with monist tendencies) inclines me.

One way of explaining Vico's confusing mix of insights and nonsense
is that he was proposing a genuinely *New Science,* and like all sciences,
it has to start somewhere. The first results will necessarily be mixed.

And this is as it should be, since what it offers is not truth
but *a way to evolve* as a full participant in what it studies.
I read Vico and, through all of what seems otherworldly,
can't help but recognize that *I am one of his descendants.*

The kinship is confirmed in that *this recognition is also how Vico reads classical texts.*
I too had to begin with what I knew was caricature ("everything seems wildly wrong"),
but making this mistake enabled me to begin thinking it through.

I am especially taken by Vico's use of the term *providence* to name
a direction in evolution that benefits individual and collective—
though my version would tend to factor God out of the picture.

Now that we are increasingly having to come to terms with traffic
across the nature/culture border, we can expand Vico's definition
to include that which benefits both humans and their environments.

So it is that providence, in the most general sense, can be said to provide
the initial conditions for the emergence of complex systems such as life,
and the pathways along which such systems can sustain and evolve.

All these involve processes that optimally serve the system in question
and its components, its environment, and relations to other systems.

For Vico, providence does not prescribe a *necessary* direction
of history or evolution but offers something like ongoing *opportunities*
that, at any given time in the human world, have their champions.

I pick up the term *providence* because we seem to lack a paradigm
for systematically identifying and pursuing such opportunities.

For emergence and evolution, the scientific paradigm is based on *chance*,
which supposedly guarantees by sheer numbers that certain chemical soups
under certain conditions will spawn life, that certain mutations will be adaptive.

Because the chance model is so inadequate, it leaves room for its polar opposite
to keep sneaking back in: the belief that evolution is inherently progressive.
When applied to the human world, this is called "Whig history."

Whig history is still going strong in neoliberal ideology, whereby the logic of capitalism,
via the "invisible hand" of the marketplace, will produce the best of all possible worlds,
if we will only recognize that *resistance is futile*. But even the opponents of this fascism
are inclined to talk about our collective *choices* as if we had sovereign free agency.

Acting as if we had choices is another disastrously reductive model in the best of times,
but, speeding past the tipping points of climate change and peak oil, it's a real trainwreck.

Even so, many readers (I suspect) will bristle at recycling the term *providence*,
even with God factored out. Mono-rationalists tend to look down their noses
at taking something supposedly abstract and making it out to be more entity-like.

But as we've seen, systems occupy the interzone between subjects
and objects: the notion that forces are abstract or objective
is as reductively wrong as personification and subjectivization!

A much stronger objection, in my view, is that various horrors—
genocides, eugenics, crusades, and jihads—can be perpetrated
in the name of something that looks a lot like providence.

This provides all the more urgent reason *to learn how systematically
to distinguish providential opportunities from their many impostors.*

If I'm reading Vico right, he suggests that champions of providence
might eventually prevail—even as decay, rampant individualism,
and the War of All against All seem to be getting the upper hand.

Although I'd downgrade *prevail* to *survive*,
I still regard myself as a committed optimist
(and hopefully not in the sense that a person
may be committed *to a mental institution*).

I think I may even agree with Vico that the champions of providence—including
scientists, activists, mystics and mythmakers—form a group that *might cohere*
eventually into something more like a religion than a science or a politics.

Of course, it would have to be all of these, *each translucent in the others*.

CHAPTER 6

God 3.5B (A Nearsighted Evolutionary Panorama)

> The … focus should be on developmental logic,
> rather than developmental details.
> > P.Z. MEYERS

∴

> … One cannot take too much care
> in handling philosophical mistakes,
> they contain so much truth.
> > WITTGENSTEIN

∴

Dear Reader, this chapter comprises a more systematic attempt *to find* (or *to found*) the principles of an anarchic and entifying religion in biology, ecology, and evolution via systems theory. If you found previous chapters suggestive but impressionistic—a series of *beginnings* and ground-clearing gestures—then here's finally *the middle*.

Here, slowly and meticulously so you can see that I have nothing up my sleeve, with longer words (sometimes so long that even I don't know what they mean), longer sentences and stanzas, I attempt to pull religion out of a scientific hat.

If that doesn't sound pleasurable (the word *systematic* does make it seem a bit of a slog), I suggest you skip it and go on to the next chapter—or you could read just the asides (there's one on the TV series *Law and Order*, one on the trickster god Monkey, and one on the evolution of the clitoris) to get through the middle and on to a series of codas.

1 Salience to Sentience

Starting in the middle (as we must), let's say we are lowly creatures, a few steps up from being involuted little webs of chemical reactions, (as are all living things), and that the presence of a certain chemical, not useful to us in itself, tends to indicate that food is nearby.

When I first encounter the chemical, a sensory detector in me is tripped.
Since the chemical has no particular significance, this is a subtle event:
some molecules of the chemical interacting with my receptor molecules.

Given the ongoing chemical chatter of receptor events, it will take much more than this
to appear on my radar at all, and even more, in turn, to tickle my fancy and to move me,
but sure enough, soon after, the cloud of other chemicals that surrounds the food itself
trips my food response mechanism, marshaling my attention and motor resources, and
causing me to emit signals that tell my nearby species-mates to join me for dinner.

The more the sequence of chemical-followed-by-food happens,
the more my sensory and response mechanisms get wired together
and the more we all will be coordinated (socially wired together)
by signaling triggered by the chemical and its new significance.

(By the way, note that the cells involved aren't necessarily neurons,
which evolve late in the game, but may be simple reflex chains—
and that the wirings and triggerings would have been made possible
by genetics, which selectively shape and transmit these possibilities
from generation to generation. What kinds of creatures are we, here,
in this hypothetical account, and where in our evolution? I am trying,
by the sketchiness of this account, to keep these questions open.)

The chemical now *means* food, triggering anticipation: when I smell it,
I send the message, and we all start looking around excitedly for food.

What may eventually be our noses have also gained new significance
in the process, and there is evolutionary incentive for them to grow,
along with the wiring that connects them to our food responses.

If the wiring is not yet a nervous system, much less a brain,
this process may send it a step on the way to becoming a brain.
Our environment harbors all kinds of chemicals—indicators
of food, potential mates, competitors, predators, prey, poisons—
so there are big survival benefits for heightened discernment.

And because our chemical environment changes over time
(and our predators and prey and competitors get smarter too),
there is selective pressure for an ever-increasing capacity to learn—
for increasing the plasticity of our triggerings, wirings, and processings.

But for the moment, since more of the chemical produces more receptor events, like a Geiger counter that chirps faster the closer it comes to a radioactive source, even this basic apparatus, which only registers signal intensity, orients me to *spacetime proximity*, enabling me to swim *up* the chemical gradient towards the food, to close in and gobble it up when I reach *the here and now*: I have begun selectively to map the world, suffused with information as it is for me, and together with it to think and feel my way along a virtual spacetime landscape of *meaning*.

After generations, we have better noses and growing clumps of wiring that act as switchboards between our noses and our response systems. These clumps could be called *transductors*, because they translate an incoming environmental signal (chemical, in this case) into new forms (chemical and electrical) specific to the system in question (that is, to *me* and to *us*). Of course, the signals have to differ from each other, enabling them to be routed properly: to put it abstractly, call one a *dot* and another a *dash*. We have taken another step toward consciousness and language without even having brains: we have begun to operate (internally) with an emerging set of transductive *tokens*, and the set of chemical or other tokens we exchange with each other (externally, or as one might say, *socially* or even *culturally*) will also be growing, and our processing of them becoming more and more sophisticated.

The clump of wiring in which the tokens are processed is necessarily growing with them, and since it will benefit us greatly to be able to process them better, we are developing a new subsystem: as my transductive subsystem grows, so grows the super-subsystem that coordinates and monitors the processing; call it an *operator* of the switchboard. As the switchboard operator gets better at manipulating the tokens, it will also need to get better at regulating *itself*, damping down the potentially overwhelming and repercussive reaction cascades to which systems are subject and, just as important, trying to stay alert and ready during lulls in incoming activity, going to the break room to take a nap, and so on.

As the operator gets better at self-regulation, so grows the super-super-subsystem that might come to warrant the name *consciousness*. The individual manifestation of consciousness (emerging from the ways our brains are wired to themselves) is linked with its collective manifestation (emerging from the ways we are wired to each other). The clump is growing clumps of its own—involutions, inflorescences—and it seems more and more to have an *inside* as the tokens and their manipulations begin to take on a life of their own: a whole little terraformation happening in there.

Soon we'll be able to manipulate the tokens speculatively, heuristically—to *hypothesize* by running "if-then" simulations in which we experiment on a wide range of tokens (and now, meta-tokens), including one we now call "myself," one we call "you," one we call "the result of the interaction between us and our tokens of each other" ("our relationship"), and so on—maybe not quite *ad infinitum*, but close enough.

There you have it, in a few short paragraphs:
my account of how I went from *chemical clusters*
to *consciousness*—from *salience* to *sentience*.
It's been quite a ride! But have I come so far?

2   Primordial Systemhood Membership Narratives

The above is a speculative account based on my *know-how*
(about how systems "lift themselves up by their bootstraps")
but with only a little *knowledge* of biology and evolution.

Since writing it, I've come to learn more about scientific attempts
to reconstruct the same story, in particular via the remarkable book
The Ancient Origins of Consciousness by neuroscientist Todd Feinberg
and biologist Jon Mallatt, which I will consider at the end of this chapter.

I take Feinberg and Mallatt's account to be correct in all ways,
though I have no scientific standing to corroborate or contest it.
I also consider it to be wrong in almost every single word—
in the choice of subjects, verbs, and objects in its key sentences—
and I hope to show how this wrongness (which, as you will see,
necessarily also afflicts my own narrative and all narratives)
points toward the kinship of language with living systems.

Since telling my little story, I've learned that there is still no consensus
as to whether vision or smell led toward the acquisition of consciousness.
In any case, to simplify the story, I reduced multiple senses down to one,
sacrificing detail for fundamental principles. I have overemphasized
the early role of communication among species-mates (though obviously
this became important at some point on the way; it's a question of timing).
I underemphasized the role of predators—and of environments
and other kinds of creatures generally—as drivers of the process:

chalk it up to my hedonistic temperament or even an American tendency
to identify "pursuit of happiness" as our prime mover. And I've omitted
mentioning anything about the role of genetics (more on this below).

So what, if anything, can a narrative that simplifies and omits so much,
that gerrymanders the cast of characters—and plays so fast and loose
with the chronology—be said to get *right*? This goes beyond structuralist claims
that all stories are reconfigurations of the same elements, all linguistic variations
permutations of a universal grammar. This goes beyond stories and even language
to reach for something like a *primordial systemhood* characterized by *someness*.

Someness is a way of being plural and singular at the same time, a way
of splitting the difference between individuals and a collective—something
not embodied by Dubhe, Merak, Phecda, Megrez, Alioth, Mizar, and Alkaid
(as stars of the Big Dipper, they are *Plural Singulars,* not a *Primordial System*)
but ably performed by Metaspriggina, Herpetogaster, Wiwaxia, and Anomalocaris
(creatures of an old ecosystem) or by cerebrum, tectum, cerebellum, and medulla.

Star members of a constellation aren't altered in nonlinear loops
by their membership; co-evolving lineages of creatures
and their organs, and the organs of their organs, are.
Altered. By their membership. *This is their story.*

> ASIDE: Law and Order
>
> I took the "this is their story" line from the old TV series
> *Law and Order*, which (as the opening voice-over tells us)
> offers itself as an account of "the police, who investigate crime,
> and the district attorneys, who prosecute the offenders."
>
> There is a glaring problem with this opening line,
> closely related to problems with narratives generally,
> including the evolutionary narrative rehearsed here.
>
> District attorneys don't prosecute *offenders* but those they *accuse*
> of offenses; a small thing semantically, but it runs fully counter
> to the supposed first principle of "innocent until proven guilty,"
> showing how fully the show takes the perspective of The Law.

In a single statement, it reveals how every narrative spun by the show—all dynamics of plot and character—are organized around the presumption of guilt.

The statement's category mistake is in treating crime and offenders as referential objects rather than as performative coproductions of the legal system (which simultaneously finds and invents them, as all systems do with their components). In logic, this is known as *begging the question* (i.e., assuming what is meant to be proven), but it also marks the circular causality by which all systems operate.

The question of how legal systems *produce* crime by *criminalization* (that is, by defining certain activities as crimes) is, in fact, sometimes represented on the show, but predictably, only by cases in which DAs find creative legal ways to hold bad guys responsible (that is, cases where we already know that the bad guys are bad), *not* by exposing the imperative to classify certain behaviors, especially when done by certain categories of people, as crimes.

The difficulty of rewriting the statement shows how the network of meanings in which it is embedded exerts its inertial force. Even if we leave aside the complexities of criminalization and try a minimal solution, keeping the syntax (as lazy rewriters often do), we can't say "the district attorneys, who prosecute those who will have been proven to be offenders if the prosecution is successful": if nothing else, the retroactive convolutions of future perfect verbs are likely to lose our viewers. But if, instead, we more simply say "the district attorneys, who prosecute the accused," we come close to collapsing the statement into a tautology, into something like "the district attorneys, who prosecute those whom they prosecute."

Only a focus on system dynamics can help us generate a viable rewrite: "the police and the DAs, whose jobs are to hold someone responsible." It's simple and to-the-point, revealing that the law-and-order mandate is not to assess guilt or innocence but *to hold individuals responsible* (acting as damage control in focusing on individuals instead of systems).

The law does this by *narrativizing according to realist conventions of characters, causality and plot.* This reveal calls attention to the show and to *law and order itself* as a plot-and-character-producing machine.

> The same realist conventions of plot, character, and causality
> afflict the story of evolution. Little tinkerings will not yield
> a viable rewrite. You have to squint. Back up. Reassemble.

3   The Myth of the Acquisition-of-Consciousness Moment

In *Terminator 2: Judgment Day*, the world-dominating artificial intelligence system known as Skynet is activated on August 4th, 1997; "begins to learn at a geometric rate"; "becomes self-aware at 2:14 a.m. Eastern time, August 29th"; and starts wreaking havoc.

In the film's precursor *2001: A Space Odyssey*, acquisition of consciousness can't be pinned down so precisely, but it has clearly passed a tipping point when the computer tells an astronaut, "I'm sorry, Dave. I'm afraid I can't do that" and, "this mission is too important for me to allow you to jeopardize it."

The anxiety that our technologies will surpass and displace us derives from our conviction—mistaken in the first place— that they are our tools and we are their masters (a mistake with roots in the subject/object distinction in our language).

Things look much different when you start instead from the recognition that we do not simply control our emerging, pervasive, powerful metasystems (such as languages, cultures, societies, economies, ecologies, and technologies): *we coevolve with them, just as all subsystems, systems and metasystems do.*

Evolving humans—our layered, complex, growing brains and social groups— coevolved with layered, complex, and growing language. None of these could have emerged and evolved without the others. Each is the tool, each the technology, the mediator, the subject and object of the others.

One of the most notorious denials of anything but individual human agency is the NRA's assertion that "guns don't kill people, *people kill people.*" Expanding this disastrous tunnel vision requires recognizing that we operate on a more level playing field with a range of other quasi-subjects/quasi-objects, each of which—like us—has *partial* agency: people are killed by people, and by guns, but also by ideology, by toxic masculinity and white supremacism— and yes, by the imperialist hubris that we are masters of our machines.

As all of this goes to show, the anxieties that our technologies will master us are not baseless after all but part of a performative or self-fulfilling prophecy *as long as we keep believing we are the masters*. We continually adjust ourselves (via Hegel's *master/slave dialectic*) to interface with our enslaved technologies, allowing them to enslave us in turn—but without the revolts and wars depicted in films like *2001*, *Terminator*, and *The Matrix*.

It is apt that a nation built on the *idea* of freedom and the *institution* of slavery—and still haunted by both—would lead the way into virtual enslavement by technologies that we have always believed are our servants.

4 Use of the Terms *I* and *We*

A key to my evolutionary narrative can be found in how I use the words *I* and *we*.

The *I* of my story is what I call the *Phenomenological I* (although philosophers may consider this a perverse or ironic usage) by which I mean, it posits ways my own experience resonates with others, where "my own experience" is a *proposition* rather than a report (which is why one encounters it most often in philosophical writing).

This proposition of the resonance of my experience with others—others who may be very different—is sometimes valued as empathy and recognition. Then again, it is just as often discounted by the charge that it is mere projection or anthropomorphizing—or, yet again, validated by assertions of actual kinship (whereby, for example, the bases of our brains bear family resemblances with the brains of all vertebrates). For the moment, anyway, I want simply to recognize this particular use of *I* and acknowledge these complexities.

> Chuang Tzu said, "See how the minnows
> come out and dart around where they please!
> That's what fish really enjoy!"
> Hui Tzu said, "You're not a fish—
> how do you know what fish enjoy?"
> Chuang Tzu said, "You're not I, so
> how do you know I don't know what fish enjoy?"

In fact, when I wrote my little story above, I was focused on its content, so I didn't notice until after the fact that *the place from which I speak* (my *I* and *we*) *has been moved* by the emerging-and-evolving systems

I've been studying for so long (and by therapy; by long histories of love, friendship, and collaboration; by roles in various organizations, and so on).

Because I was trained as a literary theorist, I can read my own writing, notice that the grammar has shifted, and learn something from it.

A subconscious grammar that can change and evolve operates as we speak and write. Grammar typically operates without our being conscious of the rules we are following, but self-monitoring (metacognition) can sometimes detect it and figure out its rules. Literary criticism—like other reflexive knowledges—has *made my brain more plural.*

If I can learn something without knowing it, I can also learn that I have learned it. Because the learning changes what I call *I,* I would not have been able to notice it at all unless my brain were plural, unless some parts of it stood outside what is called *I.*

It's misleading to use *I* and *we* to encompass thousands of generations of evolution, as if they'd assembled and elected human consciousness (*my* consciousness, no less) as their spokesperson: *Royal We, indeed!* But notice again that it was already misleading—in exactly the same way— to use the everyday, grammatical *I* to unify the radically heterogeneous choreographed menagerie of systems that make up each of us.

I am speaking *experimentally.* What do we have to learn in order to start using the leveled *I* of thinking systems, and what happens when we do? What can we think *into* when we learn to speak intersubjectively/objectively, standing here on the bridge between and with amoebas, automata, and angels?

(And by the way, this metaphor may be a bit easier to understand if you live in New York City, where there's a *Triborough* Bridge.)

5   **Cis-Systems → Trans-Systems**

Subsystems, systems, and metasystems emerge together: wherever you find one, there will be the others, which is why *systems are always* nested—*in the middle of other systems.* Systems are only *sub* and *meta* in relation to each other.

What we call a *system* is a metasystem to its subsystems, subsystem to its metasystems, and so on, for each of them. We have the adjectival markers *sub* and *meta* but not for a system itself, which is why we resort to calling it a *system itself.*

I suggest the adjective *cis*, meaning *this side of*,
as opposed to *trans*, meaning *the other side of*.

Cisgendered people are not called upon to explain why their gender matches
the sex they were assigned at birth because an elaborate sex-gender system
operates to *naturalize* their particular constellation of sex and gender features.
Just so, cis-systems can operate in ignorance of their sub- and metasystems.

But naming cis-systems is a step on a slippery slope to recognizing
that *all* systems are *subcismetasystems*, which is just a way of saying
there is no such thing as a cis-system. All systems are trans-systems.
Having invented the word enables us quickly to set it aside.

The complexity, the simultaneous singularity and plurality, openness and closure
of a system, and of our subjecthood and objecthood, is not only a local inflorescence,
an island nation with a king and a castle and a flag, defended against outer chaos.
This blessed plot, this earth, this realm is not *our sceptred isle*
but a sprawling and unsceptred archipelago.

See what just happened?

There's been a leveling, and what was a hierarchical set of levels
became a network. When I say "the whole is part of the parts,"
it's how Blake meant "to see a world in a grain of sand" and why,
in ecology, the humblest cyanobacteria are Most Valuable Players
in life on earth alongside human beings, the drama queens
of the Anthropocene—or as Bob Marley put it:

> Babylon throne gone down, gone down,
> Babylon throne gone down.

6   First Caveat: Characters

As is widely known, narratives tend to overemphasize discrete characters
at the expense of understanding more impersonal sets of forces, exaggerating

the difference between the two. This means that, from the beginning, my story
(like systems theory itself) overemphasized system-environment distinctions.

As the story indicated, though, systems grow and internally differentiate
along with growing and differentiating traffic with their environments

and with each other. This means that, as they are becoming ecosystems
and environments unto themselves, they must also be participating
in building and being built by ecosystems and environments around them.

For example, chemical signaling goes on among my subsystems, among
me and my fellow creatures, and among us and our environments.
These three kinds of coordination are themselves coordinated, meaning
systems are networks of signals that are nodes in larger networks of signals.
You can't have one without the others, so they emerge and evolve in tandem.
(This is why *semiotics*, the study of signs—if developed to its full reach—
would necessarily have included geosemiotics, zoosemiotics, and so on.)

In biology and evolution, we should try to avoid *exactly what other writers of stories should avoid*: the hackneyed, system-versus-environment account of a heroic creature—
the story of life beating the odds and "climbing Mount Improbable" (as Dawkins put it).

It may be easier to recognize this reductive use of character in fiction, where an editor
can sometimes toss such stories on the reject pile after reading the first sentence:
"The microbe tensed his flagellum as he turned to face the raging intensity gradient."

7   Plant and Animal Intelligence

Notice that I cast my protagonist as a creature who can *move*, indulging my own bias
as an animal rather than a rooted plant. Plants have their own sophisticated repertoire
of responses to their worlds (their own intelligences, individually and collectively),
but an increased ability to move under one's own power, and the kinds of interactivity
that come with it, open up new developmental pathways.

In my little story, for example, the development of my nose is,
surprisingly, spurred on by my being able to move, and vice versa:
it wouldn't help me to register smells if I couldn't also respond,
such as by moving toward or away from them (though plants
respond and adjust, too, while staying rooted)—and conversely,
why develop an ability to move myself unless I can register
what it would behoove me to move toward or away from?

It may be the case that developing a mind that grasps and manipulates concepts,
as humans do, is contingent on having grasping and manipulating hands:
comprehension comes from *prehension*. In any case, only a creature dependent
on vision and prehension would have considered it a cognitive achievement,
as I did above, to get conceptual *leverage* on something otherwise too large

and sprawling to notice (which involves backing up far enough to *get perspective*, and then *getting my arms around it*). This is not at all to say that only creatures with prehensile hands are intelligent, but that our *rootless* and *handling* intelligence is different than that of plants—dramatically but not *radically* different, rooted still in a primordial intelligence of which plants and animals are interweaving offshoots.

In an early merger, when we were all single-celled creatures, our common ancestors domesticated a microbe that would become the mitochondria that power our cells. Plants and animals started to go their separate ways about 1.6 billion years ago. The ones that adopted cyanobacteria became the plants. The fact that the branches of the tree of life can converge via these mergers—as well as diverge—goes to show that, at its base, it's not really a tree but something more rhizomic: a network.

Plants adapted their new proteins (called *phytochromes*) as photoreceptors to detect red- and far-red-spectrum light and regulate flowering and other rhythms.

As in my evolutionary memoir, the path taken by plants involves the cross-coordination of an evolving constellation of relationships: (1) the relation of plants to other creatures (as via the transfer from cyanobacteria), (2) the relation between and among subsystems in single individuals (the use of phytochromes in self-regulation), (3) among each other (via flowering and pollination), and (4) with the world at large (such as by their relation to particular light wavelengths). Since we animals didn't capture and domesticate cyanobacteria, we don't have phytochromes, aren't much attuned to far-red light, and only flower metaphorically. But our divergence from plants made possible our weaving-together with them—our coevolution, our ongoing *fraying and braiding*— which continues to shape us both.

No wonder, then, given our different paths, that plant intelligence seems so foreign to us. No need to go far afield to answer the old question of whether we could recognize forms of intelligent life on other planets: we often don't recognize them on our own!

8 Learning and Evolution from a Systems Perspective

My evolutionary/systemic use of *I* and *we* tends to complicate the distinction between *learning* and *evolution*. Learning is usually defined as taking place in a single individual and depending on the reconfigurative plasticity of neurons and other components, but how do you factor in the collective bodies of knowledge and know-hows that are what we learn—*and are themselves evolving*? *Evolution* is usually defined as requiring many generations and as being routed

through a reconfigurable genome (as when a mutation gives me a more sensitive nose,
better enabling me to survive and have offspring, who ensure that the mutation spreads).

The capacity to learn emerged through evolution, as did evolvability itself.
The circularity of "evolvability evolved through evolution" is not a mistake.
It is an example of how language presses against the constraints of its linearity
to point to the nonlinear and self-reinforcing selection of developmental pathways.

In my story, the capacity to learn—to rewire my sensory-motor connections
via interaction with my environment—is part of a self-reinforcing pathway
that incentivizes increasing my capacity to learn (by continuing to develop
a better nose and legs and brain): thus we move into the systems perspective,
in which evolving legs and locomotion are part of our learning apparatus,
and learning and evolution are players on the same leveled playing field.

What evolves in evolution are constellations of relationships. The constellation
of me and my subsystems and my species-mates, us and our environments,
also necessarily includes *us and our genes*. The addition of the genomic level
to my story doesn't change its systematicity. A gene is no freer or more constrained,
no more determined or determining, no more or less its own entity
than any of the other components. Whatever the gene does has to be approved
by its individual carriers and by cospecies creatures and their ecosystem,
and if it tries something any of them don't like (if they'd rather die than adopt it),
it will be vetoed, just as whatever the creature does has to be *within the parameters*
set by the gene and the environment, and so on. All of this together is *selection*.

Within the parameters does not mean *entirely shaped by* them.
A parameter always leaves wiggle room because the dimension
in which it operates—the dimension its operation *establishes*—
is necessarily specific (which is what makes it a *parameter*),
but those subject to it are not usually one-dimensional.

Even the so-called laws of physics, which seem unrelentingly
to demand the blind obedience of everything in the universe,
don't care if you are a rock, a star, a rock star, or a grove of quaking aspen.
It's a pretty good deal: I've agreed to accept the laws of physics, and in return
they turn a blind eye and allow me to read and write and dream and fall in love.

In fact, physicists are defined by their discontent
with coloring inside the lines of physical laws.

They're always pushing at every boundary, understanding
every interaction as a power struggle in which to assess
what they can get away with, like mischevious children.
Atomic nuclei must be made of protons and neutrons?
Well, how do you like THIS, then? SMASH! BANG!

Likewise, as long as reproduction goes on, there's a lot we can get away with,
and we've cleverly figured out ways of increasing the "realm of freedom"
within the otherwise draconian regime of DNA. Meanwhile, DNA has also
exercised its freedoms. While we DNA carriers have refused to pass along
any DNA that fully compromises our viability (that is, we refuse by dying
before we can pass it along), DNA also tirelessly keeps pushing
at every boundary, never minding if we veto 999 out of 1,000 of its ideas,
always trying to sneak in some little elaboration or work-around.
Just like water trickling down a rocky slope or flames leaping upward,
it is always in the process of exploring many weaving pathways.

The planet and the solar system have a say, too: they let DNA get away
with dinosaurs for some millions of years, but then (like the old god
who got mad and destroyed all he had created) sent a giant meteor
(*How do you like this? SMASH!*). Undaunted, DNA
set to work immediately on millions of new projects.

Everything in our environment does not belong to our local systems,
but it's complicated. Humans and meteors both belong to the solar system,
so we both have to march to its tune, and this includes us being subject to meteor visits.
The evolution of life on earth has been impacted by major comet and meteor strikes
at various points in its history, and it will again. It is an asymmetric relationship:
we seem to be more subject to meteors than they are to us, but meteors *are* also subject
to life on earth when they enter our atmosphere, breathed out by bacteria and plants—
and able to burn up most meteors that get too close. Perhaps we humans
may one day expand the planetary meteor shield built by our friends, the plants.

The point here, again, is that our environments include our meta- and subsystems
as well as *things partially outside our systems*, and we've only come as far as we have
either by staying out of their way as much as possible or—insofar as they encroach
on us or we encroach on them—by making them dance to our tune, to spin around
in our own little eddy, as we have to dance and spin to theirs.

It remains to be seen what will become of the human eddy in the network of interactions
built by plants and microorganisms. Are we a cancer, or a parasite that kills its host?

Like the cyanobacteria and plants—who, in the process of oxygenating the atmosphere,
enaged in a planet-wide genocide of anerobic organisms (some few of which managed
to survive in deep, dark hiding places)—can we, in the wake of planet-wide destruction,
also build something more sustainable?

Stay tuned—but as the anerobe said, I'm not holding my breath.

9 Second Caveat: Plot

Narratives tend to overemphasize sequential plot, making aspects
of my Chemicals-to-Consciousness story particularly misleading.
I've put the "if-then" simulations at the end, making consciousness
the pinnacle of Mount Improbable, reached only at the conclusion.

Consciousness, here (more like what Feinberg and Mallatt call *self-conciousness*),
is built on the ability to manipulate conceptual tokens in their own right,
loosened from their ties to things in the world and to various reactive responses to them.
It is a kind of further withdrawal from and a more intensive interaction with the world,
but so was the living system as it began to emerge in the first place, when there was just
a set of chemicals coming into relation with each other (a primitive metabolism)
as well as to chemicals in their environment. This is a basic feature of any system:
it selects the differences that will make a difference to it and, in the process,
comanufactures its own components.

By the way, differences-that-make-a-difference was how structuralist theory
described language and cultural systems, but it applies just as well
to the way plants use phytochromes or DNA uses the nucleic acids
cytosine, guanine, adenine, and thymine (its four-letter chemical language).
This wide application works to undermine the exceptionalism
by which we tend to think about thinking, language, and consciousness.

I smell the chemical and am triggered (and by emitting signals,
trigger my fellow creatures) to begin looking around for the food.
But let's say a predator has learned or evolved to emit the chemical as well, so
when we're lured and come nosing around, it gobbles up as many of us as it can.
Can we can learn and evolve quickly enough to avoid extinction?

Maybe those of us *least sensitive* to the chemical and other signals
will flourish while the rest get gobbled up, at least in the short term.
Conversely, some of us could develop even more sensitive noses

to detect subtle differences between the food version of the chemical
and the predator version—or we could develop an ambivalent response
better suited to the new food-*or*-danger meaning of the smell.
We could also develop *other senses* (note to self: seems like a good idea!)
to enable us to triangulate among various sensory data streams
and respond accordingly. In all of these cases but one, to evolve
is to go *in the direction of meaning*, spatially, in the moment (running
toward or away from), or in the long run, by negotiating our meaning
to our fellow creatures and theirs to us. The exception is the scenario
in which the least sensitive among us come to dominate:
they can't detect the food and aren't lured or repelled by the predator—
neither has meaning. But this is unlikely to be a winning longterm strategy.

What I gained in developing a nose can be called an enhanced capacity for futurity:
early on I started to *anticipate*; first, simply because one subsystem got wired to
trigger another, and much later, when I started to run "if-then" scenarios
(sometimes known as *thinking*) in which I imagined multiple possible futures—
though even in the first instance, myself and my fellow creatures collectively
could be said to have been *computing*. Somewhere along the way I acquired
something like *emotion*, before I had a brain and only had a primitive nose
that worked to fire up my food responses.

I've always liked psychoanalyst André Green's definition of emotion
as "the anticipation of a meeting between the subject's body and another's body,
real or imagined." *Anticipation* is a tricky word, involving, as it does, anything
from simple reflex (a mechanical process that might even be within the repertoire
of a modestly complex set of chemicals) to full-fledged foreknowledge and cognition
(which require self-consciousness). This indeterminacy cannot be resolved
if only we knew more; quite the opposite. The better the resolution, the more
we discern not the distinction but *the fractal interpenetration* of the two states.

We're more like the supposedly simple sunflower that turns toward the sun than we think.
The capacity for futurity, for exploring forward into the "adjacent possible," is not only
definitive for all systems but may be a way of thinking about the nature of time itself,
not a one-dimensional and one-directional arrow but an ongoing fraying and braiding.

At what point in evolution does emotion as such arise,
and how is it distributed among individuals and collectives?
Again the most telling examples are the most ambiguous.

Bees form a superorganism that crowd-sources decisions among its members. They return to the hive and their waggle dances indicate where food is to be found. Bees at the hive are triggered to follow when an information threshold is reached. They follow the strongest signal: the most numerous and intense waggle dances. As in my chemicals-to-consciousness story, the sending out, receiving and processing of multiple signals—and the competition among them to move us—happens both among fellow bees socially (externally) and inside the brain/body of an individual (internally). It is easy to see in this example that the signals themselves can evolve (here, into the modestly complex waggle dance) along with the collective neural net and with the brain that processes them, and that this is not fundamentally different from the evolving sophistication in processing sensory stimuli from the environment. The competition between plural impulses is also at the core of what we call *emotion*.

Emotion is inherently complex and frictional. We can confirm at least that one old saw about narrative, that it needs conflict to make it move.

You could emphasize the mechanical, reflex nature of this process—for example, the way light falling on a leaf or stem might mechanically twist the plant toward the sun. But look closer and the mechanisms tend to be more fractal, *more Rube Goldberg* than they may have first appeared. Even plant-twisting involves transduction and the activation of phytochromes to receive light, which is why rocks don't twist toward light. And again, as always, let's take care to avoid excessively aggrandizing the complexity of our fellow living creatures. In fact, photosensitivity is a property of various inorganic chemicals, so we also have to complicate any assertion such as "rocks don't twist toward light": what about silver halides, used before digital photography?

In *Ancient Origins of Consciousness*, Feinberg and Mallatt recognize the continuum between the temporal dimension of sensory, cognitive, and emotional experience, but their focus is on *the past*:

> Smells tell of the past. Because smell has this unique time dimension, as soon as olfaction appeared, natural selection linked it to memory: from this smell, I remember and recognize what odor-emitting object used to be here and may still be nearby, so I can know how to avoid or approach it. The intimacy of this relationship explains why memory and smell structures evolved so near one another in the incipient telencephalon of the ancient vertebrate brain. (84)

The word *unique* is a bit sloppy: hearing and vision also have a time dimension because the nearer in time (as in space) we get to something, the louder and larger it gets. As I swim up the smell gradient (toward greater chemical concentration and thus more intense smell), I get closer in space and time to the smell's source. Fainter smell means more time and space separate me from its source; distance and simultaneity are relative and defined by *intensity*.

The assertion that "as soon as olfaction emerged, natural selection linked it to memory" also seems to be fundamentally wrong in another way, though I suspect the problem is in the selection and ordering of subjects, verbs and objects. It implies that memory was present, olfaction appeared, and then something like a third entity, natural selection (to which the sentence provisionally assigns subjective agency), linked the two. But smell and memory and selection must have arisen *together*. I define *selection* here, again, as an emergent property of the interaction of all the components of the system (the creature and its prey and predators, their genes, their wired-together sensory and cognitive and emotional subsystems); the way *a set of futures* is selected.

The passage also contains a remarkable slippage. The assertion "smells tell of the past" seems at first to refer to the *object*: the thing being smelled, even if not visible, must have been in the area in the recent past. But the assertion also refers to the linkage of smell with memory, an entirely different and *subjective* phenomenon: I remember that I have experienced a particular smell in the past and with what it is associated. And finally—this time as part of a *transindividual* and historical phenomenon— smell links us with our ancestors stretching back into the distant past.
When Feinberg and Mallatt use the construction "from this smell, I remember" (what I called the *Phenomenological I* above), they place themselves in the position of an ancestral creature—or rather, the *I* of the sentence is ambiguous enough to apply to any creature with olfaction and memory, from the first vertebrates—to the authors.

As Lacan put it, "Meaning indicates the direction in which it fails." Rather than reject their slippage as a logical problem, I suggest that, even as Feinberg and Mallatt fail to embrace the ways language embodies the process it purports merely to describe, their language points toward nonlinear and mutually co-constructing relationships among objective, subjective, and evolutionary/transindividual meanings.

10 Especially Informative Postinfundibular Amphioxus Hypothalamus Neuropile

> "It seems very pretty," she said when she had finished it, "but
> It's rather hard to understand!" (You see she didn't like to confess,

even to herself, that she couldn't make it out at all.) "Somehow it seems
to fill my head with ideas—only I don't exactly know what they are!"
> ALICE, after reading the poem "Jabberwocky"

I loved this sentence from *Ancient Origins of Consciousness* the moment I read it:
> Especially informative is the postinfundibular neuropile
> of the amphioxus hypothalamus (figure 3.5B).

I loved the agglomeration of learned words—the big pile of neuro-words,
the neuro-pile of big words, all the more poetic because I could read them seamlessly
without knowing what they all meant, and because, even so, I got the general idea:
something anatomical (something structural and spatial) bears the traces
of the evolutionary processes (something temporal and potentially narratable)
via which it has been assembled. In other words, the sentence proposes a relationship
between a *figure* and a *narrative*. Of course, the figure would have to be one that,
like all systems, is in the middle of a modestly complex series: *call it Figure 3.5B*.
And because my cognitive style is figurative rather than narrative (I call it a *style*
neither to negate nor aggrandize it but to explain my dissatisfaction with narratives),
I focus on how the dynamic relationship between figure and narrative is constantly
being reembodied *into figures*. The ongoing reembodiment *is* evolution.

Think a little about "especially informative"—the easiest bit of the sentence—
and you realize that it's shorthand for the entire scientific enterprise.

How do you get from a dissected *amphioxus hypothalamus* on the table in front of you
(don't worry, we'll get around to exactly what an amphioxus is; I didn't know, either)
to the epic story of brain evolution? The sentence suggests that a chorus of brains
of dissected creatures might just sit up in their petri dishes and begin verbose expositions
(after all, they are *especially* informative), placing their life stories in relation to the lives
of ancestors and descendants. In fact, there's a website called *Children of Amphioxus*—
nice title for a long-running soap opera whose cast has been replaced many times over!

Of course, it is not the amphioxus but the *scientists* who tell the stories.
The sentence mobilizes the foundational myth of the scientific enterprise:
the notion that Nature speaks directly through science and scientists.
In any case, you must spend generations trying to pose the questions properly
(an activity that includes theorizing and experimenting), and—as in much fiction—
telling the story must be part of the story if you want to establish the continuum
of physical processes, consciousness, and self-consciousness.

So *postinfundibular amphioxus hypothalamus neuropiles* (whatever they are) are *especially informative* only to those who know how to pose the question and interpret the answer. The question that drives Feinberg and Mallatt is something like: What structural and operational complexity was necessary for consciousness, and how and when was that tipping point reached? This question is not yet posed in a way that an amphioxus could answer it.

Getting to the answer is not an empirical matter of dissecting amphioxus brains: consciousness is diabolically difficult to define; we don't know exactly how it relates to brain structure (how operation and structure are linked); brains almost never fossilize and their relationships with body parts that do fossilize are so complex that it is difficult or impossible to infer one from the other; we have only evolved versions of the brains of our ancestors from which to deduce—and so on. When in the evolution of vertebrates did the hypothalamus begin to take shape? This is getting closer to a question the amphioxus is positioned to answer—or at least, like the helper figure in a mythological quest narrative, to inform us about, *especially*.

> My great-great-grandmother used to have a neuropile,
> out behind the hypothalamus—well, we didn't call it
> a *hypothalamus* back then; it was just an ordinary place
> where some paths converged. And she used to say that,
> one day, some of our great-great-great-grandchildren
> would come nosing around, and where the old neuropile used to be,
> there would be an elaborate switching station in a bustling metropolis.

"Especially informative" is also the marker of a fruitful area for further research in an evolving theory, an "x marks the spot": *dig here*. We are still trying to find "the ancient origins of consciousness"—and along with them, the extent of our kinship with the amphioxus. If scientific consensus had been reached, "especially informative" would drop away in favor of more reductive assertions, as it does later in the book, along the lines of "the vertebrate brain evolved via *these* particular structures in *this* order and through *these* creatures." In Latour's terms, we would fall back from having to recognize "Nature Two"— including contending groups of scientists theorizing, experimenting, telling stories, and collaborating with Nature in worldmaking—and revert to "Nature One," our notion that nature is "external" to us, a set of facts independent of our stories and theories. Paradoxically, this is what we can believe only when there is broad enough consensus among our stories and theories. Objectivity is an intersubjectively emergent phenomenon; call it Nature 1.5.

But because we haven't fully worked out the details, Feinberg and Mallatt
are obliged to weave together the story of the evolution of consciousness
(in its current "through a glass darkly" version) with the story of the telling of the story
(including which scientists have contributed which part, where they agree and disagree
as to how the plot and characters develop, and how the whole thing should be framed).

My point is that this provisional or in-between understanding bears insights
that may well drop away if we move in the direction of more consensus:
a version of the situation in which the least sensitive among us come to dominate
and the potential meaning we no longer recognize comes back to bite us in the ass.

> Especially informative is the postinfundibular
> neuropile of the amphioxus hypothalamus.

Curiously, as I worked out what a *postinfundibular amphioxus hypothalamus neuropile*
might be, it only enhanced what I'd already gleaned when I didn't know the words!
This happened via the generative friction of my know-how with my limited knowledge
(that is, with my ignorance). Please bear with me for a bit of explanation.

Postinfundibular means on the spinal side of the infundibular organ,
which resides on the edge of the brain, whether inside it or not.
A *neuropile* is a synaptically dense region (*pile* derives from the Greek *pilos*,
meaning *felt*): fiberlike structures pressed into clusters that may or may not
be metaclustered into a brain. The *amphioxus* (also known as a *lancelet*)
is a small, fishlike aquatic worm: something on the way to becoming a fish,
with something on the way to becoming a brain, not quite there yet in both cases—
though you have to take into account the open question of how much
the modern and evolved amphioxus still resembles its early ancestors.
The *hypothalamus* is an ancient part of what became the brain
possessed by all vertebrates (which, in its contemporary form,
continues to perform basic metabolic and other self-regulatory functions),
though again there is the question of exactly how much modern hypothalamuses
can tell us of their primitive functions, since they have now been coevolving
with the rest of the brain over millions of years.

You may have noticed a strange, echo-chamber effect in these definitions:
every word in the sentence (and while we're at it, let's add "incipient telencephalon"
from the last section) refers to something whose leading characteristic is its ambiguity
and liminality—its *betwixt-and-betweenness*—in space and time and identity.

Again, this in-betweenness is not something to be resolved as we gain knowledge, but an ongoing condition. We are no less in-between than the amphioxus. The states we call consciousness and self-consciousness are not endpoints but more in-betweens. The in-between is not marginal to identity, but *its wellspring*. It's hard to say whether it's best to spin this as a profoundly paradoxical revelation or something so obvious it goes without saying, hidden in plain sight. What makes the centrality of the in-between so difficult is the mandate to narrativize nonlinear processes and relationships along *lines* with origins and ends, whereby consciousness began in some rudimentary form, evolved, and then, somewhere along the way, crossed a tipping point. The simplest rejoinder to this linearization is that *lines emerge from networks*. In the largest and most abstract sense, time is the emergent effect of interactions among stuff in the universe rather than a kind of preexisting empty stage upon which the stuff struts and frets. The division into subjects and verb in even the simple proposition that *creatures and their components and ecosystems evolve* makes a basic mistake: the subjects and the verb are all the same thing, the same process!

> Especially informative is the postinfundibular
> neuropile of the amphioxus hypothalamus.

I learned more about the sentence when I tried to make *a title* of it,
thereby intensifying the implosion of grammatical linearity
into the primordial systematicity from which it emerges.

To start, it's barely a sentence at all, with the little verb *is* struggling to hold aloft the unwieldy noun phrase following. Removing *is* to make it a pure noun phrase (like many titles) risks making it collapse further into a senseless jumble of words. The only thing that can keep any semblance of grammar without a verb is the unspoken set of rules for adjective order. Such rules are difficult to specify but intuitively easy to deploy (which is why, for English speakers, the noun phrase "three little yellow South American canaries" rolls trippingly off the tongue, while "South canaries American yellow little three" is gibberish).

Neuropile is the main noun; everything else modifies it. It's easy to recognize *especially* as an adverb modifying *informative* because these words have been (helpfully) marked as adverb and adjective by the inclusion of suffixes, but from there it gets harder to tell what's a noun and what's an adjective: *amphioxus* and, in my title, *hypothalamus*, are *null conversions*: nouns that function as adjectives without any change in the word, simply by being repositioned in relation to other words (as more elegantly demonstrated

by the phrases *canary yellow* and *yellow canary*). This dynamical linkage
of position and function is a big part of what grammar is and does.

Note that I am not trying to impart any particular knowledge of grammatical operations,
just as I was not trying to teach you (or to learn) the specifics of amphioxus neurology;
I need just enough specifics—the sweet spot at the edge of disciplinary knowledge—
to see clearly their evolving systematicity and their kinship. The more specialized
my knowledge (perhaps it's passed the tipping point already), the less I'll be able
to see. *This* could not have been written by a neurologist, biologist, or grammarian.

The parts of speech and parts of the nervous system (amphioxus and otherwise)
are parts of nonlinear systems, meaning that they are in dynamical relationships
with other parts. The forms of words, their grammatical functions, and word order
constitute an evolved system in which each element bears upon the others.
Like brains and their components, they could have evolved differently
and indeed *have evolved differently* (into different languages and grammars
all across the planet), and they will continue to do so.

My not knowing the words was what enabled me to see *in them*
primordial grammatical processes at work: it made the sentence less transparent.
I first saw *through them* to their referents only as through a glass darkly:
the object beyond was fuzzy; I saw the glass itself and its dusty raindrops
and, superimposed on both, my reflection. Call it the *nearsighted view*.
But in the process I recognized something about its dynamics more clearly.

If there were a universal grammar, I'd say that what I glimpsed was how
an ur-grammar morphs into specific utterances in English. But it's more.

What I saw was how something pregrammatical and fully nonlinear is linearized
into language, and beyond that, how primordial (timelike and spacelike) dimensions
and again, how trans-systems (such as creatures, their subcomponents and ecologies),
and again, how language (its verbal process, nounlike structures, adjectival features)
all keep beginning again to coalesce out of some kind of primordial someness.
The sentence does not simply *represent this,* but *enacts and participates in it.*

Of course, it matters exactly what a hypothalamus is (and in relation
to what other structures and functions), whether amphioxi have them,
in what order various structures evolved, and so on. These kinds of things
are what makes them them and us us: "differences that make a difference."

But orienting ourselves primarily to the centers of categories
(while tending to regard the edge-dwellers as means but not ends)
will yield mistakes that we can process only by reorienting to the edges,
to the betweenspace and to what *simultaneously* differentiates and relates us.

Freud linked "the antithetical meaning of primal words"—the observation
that many basic words function as their own opposites—with his observation
that oppositions (even those as basic as *yes/no*) are not part of the repertoire
of the language of the unconscious, as in dreams. One such primary word
means both *to differentiate* and *to identify*: the word is *with*, meaning both
with as we usually understand it (as when fighting side by side with someone
against a common enemy is called *fighting with someone*) but also *against*
(as when *fighting with someone* means fighting *against* them, head to head).
So: the opposition of opposition and nonopposition is a second-order one.
Primordial and antithetical *withness* is its enabling condition.
This is no grammatical fluke: antithetical withness operates physiologically.
The hormone oxytocin facilitates loving bonding (as produced by cuddling)
but it also facilitates conflict and battle, having been shown to surge,
in chimpanzees and in humans, when groups face off against other groups.

This shouldn't be too surprising: of course what I oppose defines me as much
as that with which I ally myself and identify; of course the two define each other
(every *with* is also an *against*). Because both of these are dynamical processes
defined by ongoing turbulences of complexity and conflict, "simple opposition" and
"simple identification" are always mistakes, even when they are generative mistakes.
Their generativity comes from their mistakenness, from
the way nature abhors simple oppositions and identifications.
The ongoing permutations of inside/out and with/against are *meaning*.

> ... this is how we tried to love,
> and these are the forces they had ranged against us,
> and these are the forces we had ranged within us,
> within us and against us, against us and within us.

11 The Holy Grail

Surprisingly—for sheer grandiosity—Feinberg and Mallatt
claim to have solved the "hard problem" of consciousness,
making the "mystery of consciousness" tractable

(the ironic quotation marks are theirs) by weaving together
"neurobiological, neuroevolutionary, and neurophilosophical domains."

The ironic quotation marks are a holdover from reductionist science, which aspires
to demystify everything by reducing it to well-understood material processes.
I say *holdover* because I don't think that reductionism is the main thrust
of Feinberg and Mallatt's work, but even so, *they fall back into its language.*

> ASIDE: Neuro-X
>
> Some of the self-referential force of Feinberg and Mallatt's claim
> to weave together "the neurobiological, neuroevolutionary,
> and neurophilosophical domains" is revealed if you render the claim
> in a more opaque and abstract form (as when I didn't yet understand
> their words): "to weave together the XA, XB, and XC domains"
> (where X equals *neuro* and A, B, and C are the names of disciplines).
>
> Clearly, whenever you have XA, XB, and XC, then X is, in fact,
> woven through A, B, and C. I am not saying that the forms of the words
> make the argument by themselves (that would be to use *self-referential*
> in the dismissive way it is sometimes evoked in reductionist science)
> but that the rise of the plural/singular disciplinary constellation of *neuro-x*
> participates in the plural/singular constellations it describes.

The ironic quotation marks would more appropriately go around "solving"
the hard problem—or solving the "hard" problem—because what is involved
is more like *displacing*: if you've been beating your head against the hard problem,
you should know that there's an open doorway on the other side of the room.

In fact, Feinberg and Mallatt only get as close as possible to saying they've solved it
without affirmatively stating that they have: "perhaps one reason no one has solved it
before is that it requires all three perspectives"—i.e., the neuro-disiplines—
"including what happened over half a billion years ago" (227). They also assert that

> we have philosophically bridged many of the explanatory gaps
> between natural neural processes and ontological subjectivity,
> while rooting consciousness firmly in the scientific field of biology.
>
> This could be the holy grail of consciousness studies,
> an unbroken continuity between subjectivity
> and the explainable, tractable life processes. (225)

Even with the "could be" hedging their bets, it's a funny kind of grandiosity:
we have made the greatest discovery in the history of humankind—*sort of.*

Far from downplaying the importance of the project, I want to participate in it:
the continuity of "chemicals to consciousness" and "salience to sentience" is also
that which I experienced as a breakthrough in my own thinking and writing.
But dominant strains of Western philosophy created the problem
of an unbridgeable Cartesian chasm between mind and body in the first place,
so claiming that you've discovered how to bridge it is a bit like saying
Columbus *discovered* America, where people had been living for millennia.
To put it another way, someone *may* kill the fatted calf for you when you return,
even if you left out of your own pride, but the prodigal son should be *humbled.*

I propose that the sense of solving the hard problem comes from moving
from the *twoness* of the Cartesian mind/body dichotomy to what I call *someness*
(plurality/singularity as an "antithetical primary word") or even just to *threeness*
or *fourness*. This conceptual movement is sometimes called *deconstruction.*

The "neuroarchitectonic diversity" (211) required for consciousness,
according to Feinberg and Mallatt, is a tipping point occurring
with a sensory/nervous system that has three or four levels of structural hierarchy:
"the minimum number of levels allowing consciousness in the vertebrates
is four for most of the senses but could be three for the somato-sensory pathway of fish
and amphibians" (178). A bit confusingly, these three or four layers allow
for a consciousness or self with three modes of operation they call *exteroceptive,*
interoceptive, and *affective*—and which they understand as foreshadowed
in Damasio's tripartite model of *proto-self, core-self* and *autobiographical self;*
and in Tulving's *anoetic, noetic,* and *autonoetic* consciousness.

Note how these not-quite-aligning sets of three or four echo
in Feinberg and Mallatt's characterization of their own breakthrough
as "neurobiological, neuroevolutionary, and neurophilosophical."

Just as Feinberg and Mallatt propose that amphioxi achieved consciousness
with three or four layers of structural hierarchy (something very like
the super-super-subsystem of my little story), so the scientists themselves
achieved their breakthrough—their metaconsciousness—
by thinking with the three-or-fourness of their various schemas,
by coming to terms with what I call *thinking systems,*
which as a conceptual breakthrough may be likened to the feeling one gets

the first time one smoothly juggles multiple objects: an exciting moment
for creatures like us, with eyes, prehensile hands, and a brain to match.

What are we to make of this? Have Feinberg and Mallatt simply projected
their fondness for three-or-four schemas onto the nervous systems they study?
Like Narcissus, have they fallen in love with their own image,
believing it to be another—to be *The Other*? I don't think it's that simple.
Or is there some inherent three-or-fourness that they have simply discovered,
and the three-or-fourness of their schemas simply reflect that? (Also too simple,
even by their own account.) They represent their discovery not as a matching
between theoretical schema and object but as a conceptual breakthrough,
the implication being that some sort of evolution in thought has taken place,
and somehow—in an uncanny way that goes unremarked upon in their text—
this breakthrough echoes the evolutionary tipping point it theorizes.

It's at least plausible that science had to evolve to match the complexity of its object
of study before it could understand it properly, much like a light-sensitive spot
not yet evolved into an eye may lag behind the complexity of what it is seeing until,
at some point, it overtakes it and the eye/brain wiring becomes *more* complex
than its object. It could be that the science is finally "coming of age"
as a kind of metaconsciousness and achieving a *someness* and complexity
that consciousness—both its substrate and its object of study—
had already achieved approximately 350 million years ago.

Narratives have to pick where they start and conclude, always more or less in the middle,
but if there might be some mathematically describable complexity threshold in operation
across a range of different kinds of phenomena—a place where consciousness emerges
out of clusters of chemicals, sentience from salience, and eventually science and meta-
science from sentience—a place where quantitative change can become qualitative,
where the whole becomes more than the parts, where something like a system emerges,
why stop at 350 million years ago?

It seems likely that some similar complexity threshold may have been crossed
by something as fundamental as atoms (which seem to have emerged a few minutes
after the Big Bang) or, before that, in the primordial dimensions of vibrating strings
or whatever may have coalesced into subatomic particles (in the first few seconds).
Perhaps your brain feels the narrative mandate to keep stringing these tipping points
onto a straight timeline or a spiral (the figure at the heart of the word *evolution*)—
three seconds after the big Bang for the emergence of subatomic particles,
three minutes later for atoms, three million years ago for consciousness,

three years ago for Feinberg and Mallatt's discovery of the "holy grail."
If figures like these crystallize into something you can hold in your hand
and turn this way and that (the linearized story of evolution),
perhaps you can also sense the higher dimensional manifold
for which they stand, in which we and our stories continue to be held.

12 Teamlikeness

By pressing on the wrongness of certain key concepts in Feinberg and Mallatt's account—
the way they prematurely stabilize certain concepts, making *nouns* out of them—
we can find how these wrongnesses point to what the scientists' language cannot say:
the difference between what it *says* referentially and what it *does* performatively.
The wrongnesses include insistence on *reference* as the marker of consciousness
(with *self-reference* as the even "higher" marker of self-consciousness)
and the insistence on *hierarchy*, *unity*, and *qualia*.

a   Reflex, Reference, and Self-Reference

Feinberg and Mallatt link the rise of sensory, emotional, and cognitive consciousness
to the moment when simple *reflex* (whereby the sight of you triggers a response)
gives way to *reference* (whereby I understand that you are an entity in the world
and that you have certain characteristics) and eventually enables *self-reference*
(whereby I understand and can also contemplate *myself* as an entity),
a higher form of consciousness generally known as *self-consciousness*.

As it happens, *reflexive* is another one of those words that is its own opposite.
It refers to a simple reflex—a mechanically triggered reaction, without consciousness,
usually to something outside the system in question—but also to the way a system
observes and responds *to itself*, a process that includes the "highest" faculty
at the other end of the spectrum from simple reflex: reflexive self-consciousness.

As my story was meant to show, the boundary between them is fractal;
their ongoing, simultaneous fraying and braiding is the story of our evolution.

b   Hierarchy and Unity

Say that five nodes are connected to each other, forming a cluster,
and that the cluster of five is connected to another cluster, and so on:
this is what passes for hierarchy, with the metacluster being identified
as the "higher" network. Here the concept of "entangled hierarchy"
is vital: what we now call the metanetwork may have emerged *first*

(its micronetworks may have sprung up later, starting as something like small-scale ornamental elaborations on a larger structure) or *second* (the metacluster may have arisen on the backs of the micronetworks), or some combination of both (most likely), but in any case, the salient fact is that they are likely to be interdependent: wired so intimately together that to shut either one down will take the other with it. As we have seen, subsystems, systems and metasystems (trans-systems)—creatures and genes and their fellow creatures and environments—emerge together.

In the entangled hierarchy, consciousness and self-consciousness (the *I*), often represented as a unified and unifying consciousness, is not the "highest"— a kind of Chief Executive Officer—but something more like *middle management*. If you think with the executive *I*, you're going to get some things quite wrong.

Consciousness and self-consciousness have become important members of the team they belatedly joined, but teamlikeness was its primordial and still leading feature: by its differences from the other members, the *I* did not unify the team but *enhanced its teamlikeness*.

> ASIDE: Play and Teamlikeness
>
> The mythic team of four in the 16th-century Chinese story *Journey to the West* (an account of the Tang Dynasty pilgrimage to India to bring Buddhist scriptures to China) are Monkey, the monk Tripitaka, Sandy, and Pigsy. The monk is a superego figure (he keeps everybody "on mission" and works to keep Monkey, in particular, under control); Pigsy is an incarnation of the appetites; and Sandy is more like a kind of operational principle (the strong and silent type, with his horse). Monkey is the Trickster, about whom it could be said that he pursues play not instrumentally for the sake of power but rather power for the sake of furthering play: as Gordon Burghardt (theorist of animal play) put it, play is *autotelic*.
>
> Note that I am not trying to map the four characters onto Freud's tripartite id/ego/superego schema: my point is that, as with many other such schemas, *they interact and overlap without matching up.* This is the unmistakable marker of *someness*.
>
> The necessity of the team—it being more than the sum of its parts— is the point throughout the story, as it's clear that Monkey, on balance,

may be more of a destructive force when left to his own devices.
Even so, *play is still the heart and soul of the story*, which is why
the story's alternative title is simply *Monkey*. Not an idlike life force
or an egolike reality principle or a superegolike authority figure, but
the principle of play is the closest thing to a god in the group.

Teamlikeness—and the ongoing precedence of the anarchic—is often counterintuitive
for those used to command-and-control hierarchies. It may be hard (for example)
to grasp how *removing* traffic lights and markings can make roads safer, but when
this is done, people tend to slow down, get more attentive, make more eye contact:
driving becomes a more nonlinear process, more of a negotiation in realtime.

Evolutionary sweet spots cannot be found by trying to figure out
how to strike the best deal with inevitable fascist authoritarianism
but only by exploring *how to make anarchy work even better*.

This, for me, is the philosophical imperative of anarchism.
It describes the evolution of our senses and how they cooperate
better than the notion that they invented hierarchy, rendered unity
out of diversity, and elected *I* as ruler for life.

ASIDE: Three-Ring Circuits

Many female mammals—especially those that encounter males
infrequently—release an egg only when triggered hormonally
by an intercourse-generated orgasm, reliably produced by a clitoris
located *inside* the vagina. But as our primate ancestors started
to become more social and opportunities for sexual encounters multiplied,
the female orgasm/ovulation system began to shift. Ovulation *cycles*
began to emerge, and the clitoris began to drift outward from the vagina.
A scientist, Gunter Wagner, speculates that this drift
was evolution's way of dismantling an obsolete sensor system
(phasing out "the old signal sending noise at the wrong time"),
but that's more of a "Just So Story" than a causal argument.

Evolutionary drift is more powerful than is often acknowledged:
when there is no direct selective or adaptive pressure on a feature,
genetic variations can develop and can come to have adaptive benefits
by more circuitous selection processes. As is the case with all mutation, complex
systems collaborate with accidents to make them meaningful.

Sexuality is a complex system whose components connect to each other
in multiple ways and with repercussive effects. For example, before sexual
reproduction evolved (and still for various single-celled creatures),
genetic exchange used to be *entirely separate* from reproduction
(which happened through splitting and did not require a partner).
When sex came along, it yoked reproduction to genetic exchange—
though now, through cloning and other technologies, we may be
on the verge of decoupling, or at least complicating, this connection.
Even when female orgasm was no longer required to induce ovulation,
natural selection seems to have continued to favor it, presumably
because it acts as an incentive for women to engage in sexual behavior
that might lead to reproduction. Even so, it also facilitates
assorted nonreproductive sex for pleasure—though, in so doing,
still keeping the door open for possible conception by accident,
the ace-in-the-hole evolutionary strategy of heterosex.
The prospect of orgasms sometimes gets people together,
sometimes it bonds them, sometimes they're male and female,
sometimes it leads to conception: these must be good-enough odds
for evolution, since this is where it has, in fact, taken us. Even so,
the More Sex and Less Ovulation mode seems quite unreliable
in comparison to the ovulation-by-orgasm mode:
why would evolution settle on this Russian roulette scheme,
allowing conception only, at most, five days out of thirty
(exactly the odds of one bullet in a six-shooter)?
It must have something else to recommend it.

Curiously, at least until the Renaissance in the West, it was believed
that conception required female orgasm to release female "seed"
to join with male. Even now, long after being disproven, it's still
a popular notion. Notice—the curious part—that although incorrect,
it correctly describes the evolutionarily earlier state of affairs
(whereby only orgasm could release an egg), which had to wait
another 300 years or so to be rediscovered by evolutionary scientists.
The persistence of the idea that *only pleasure is generative* is a happy one,
maybe even a feminist one, but it is also deployed in the misogynist lie
that women who get pregnant from being raped "must have enjoyed it."

Could the persistence of the physiologically counterfactual notion
come from some subtle resonance we still have with our ancestors?
The answer is circuitous, *literally*—by which I mean, to be found
in the evolved circuitousness of the sex/reproduction system.

Since pleasure, for human females, doesn't follow as automatically from intercourse as it may have when the clitoris was in the vagina, it seems that females would be more likely to select partners who give them pleasure (the opposite of rapists, more or less), which would in turn likely include a greater proportion of partners with *empathy* who care how their partner feels, if only self-servingly (that is, to the extent that each partner's pleasure benefits the other).

Cyclical ovulation also seems statistically to favor ongoing partners over one-off encounters (since those having sex regularly are, thereby, even more likely to reproduce).

These factors operate mainly through female *sexual selection* of males, though they are clearly not the only factors at work, or we'd have achieved a more thoroughgoing victory of sex-for-pleasure over sex-for-power and empaths over rapists, when it seems instead we continue to negotiate the terrain between. This is *not* to say the line is blurred (a fractal boundary is *complex, not blurred*): what rape violates is precisely *negotiation of the in-between*; rape is an act of domination outside the realm of what we understand by *sex*. (And at the other extreme, gentle empathy may not be many people's idea of sex, either.) This is not a feminist ideological definition, but an evolutionary direction whereby female choice has been wired into the sex/reproduction system.

Orgasm-triggered ovulation and the clitoris-inside-the-vagina kept reproduction a more mechanical reflex: "push button to release egg." Since then our status has changed to "it's complicated." In its new and more ambiguous role, the clitoris can't quite be reduced to a reflex ("push button to generate pleasure"), since pleasure has its own idiosyncratic ecologies: complex boundaries between pleasure and pain, the optimal mix or rhythm of direct or indirect stimulation, fast and slow, and the nonlinear/collaborative process of feeling one's way in real time along an optimal interactive path known as *play*—or as *the sweet spot*.

Notice how the developmental pathway taken by humans (the pathway that made us us) introduces more circuits of connection and causality: sex and reproduction are more socially mediated via the dynamics of coupling, sexual orientations and preferences, empathy, and play.

The notion that female pleasure is necessary for procreation,
while our evolution has rendered it physiologically wrong
at the level of the individual and the one-off encounter,
continues to be right in a collective, statistical, and evolutionary sense,
having shifted from being a physiological fact to being a social fact.

Happily, the clitoris moved to the slippery slope between inside and
outside, where it could more meaningfully participate in orchestrating
the ambiguous relationships of sex, pleasure, reproduction, and bonding.
It moved in the direction of *meaning*, which I define technically
as how the constellation of elements that compose a system
(and how these connect with each other) interacts with the way
the system interacts with other systems, subsystems, and metasystems.

Like a brain that learns by selecting and keeping neurons that have made
the most connections (and pruning those that haven't), sex has evolved
by selecting for the combination of dense and wide connections.
The densely wired nub is widely—wired and wirelessly—a player
in the three-ring circuits of pleasure, power dynamics, and reproduction.

That wasn't necessarily a forgone conclusion. Anatomical features
can also move in the direction of becoming useless appendages.
Just as the relationship between a feature and a creature can change,
the relationship between individuals of a species and the superorganism
they constitute can also change, and intelligence can be redistributed
between superorganism and individuals. In rare cases, whole species
seem to have evolved by becoming *less* sensitive: sponges no longer
have brains or nervous systems, and when sea squirts grow
from active larvae into sedentary adults anchored to the seafloor,
they radically downsize their brains.

So here's the thing: to the extent that we have a choice, let's not
go in the appendage, sponge, or sea squirt direction. Let's not
go further in the direction of ceding intelligence and meaning
to the superorganism that we form together. I'm not saying
that we *inevitably* will go in the direction in which clitorises
and human individualities will remain hubs of meaning.
I'm just saying, *let's*.

I understand this as another example of the empirical basis
of an anarchist ethics. And note that the anarchism promoted here
is an opposite of libertarianism: it requires that the system be tuned
more rather than less to optimize the balance or sweet spot
among features, creatures, and metacreatures.

c   Unity and Reference

Earlier we considered what would happen if a predator could lure us in
by emitting a smell that we associate with food. This might well apply
selective pressure for linking olfactory input with another sense—
let's say vision—that could give us competing or corroborating data:
the smell is telling us it's *either* food or danger (making it very interesting)
but then it comes into view and—*"Oh no! Run away!"* We could have unified
the competing sensory data streams via a referent—*"It's a predator!"*—
but we could also have crowd-sourced the decision among our senses (here,
it's a small crowd of two) and been emotionally moved by the strongest response,
which, in this case, despite the fascinating aroma, is prompting us to *run away*.
In either case, this is part of our coevolving relationship with our predator.

The attribution of *reference* to an external reality recognized as such,
which involves the notion that "predators are like this" (that is, that they possess
certain recognizable characteristics, such as being dangerous), is often situated
as a higher form of consciousness that now acknowledges existence of the "other."
But if it's an advance on "most of my bodily signals are telling me to run away"
(which doesn't necessarily posit an other at all), it's by virtue of making a new mistake:
projecting *my feelings* onto the other. Since this subjective projection is the very basis
of *reference*, it is no simple matter somehow to extract objective otherness,
but to the extent that it is possible, it will also introduce a new mistake,
making it a counterintuitive "higher" perspective that "you and I are co-constructed
by our relationship." People spend years in therapy to operationalize this principle—
or else remain defined by their projections, unable to recognize them as such.

But even with years of therapy, I cannot simply step out of this dynamic;
in fact, *I* am the one most defined and confined by it. The only way
I can partially *leverage* it is via my own multiplicity (the way I am not I),
the ways we are distributed among our internal and external relationships.

The emphasis on unity—both of the object and of the subject—leads to mistakes,
as here in what seems to be "obvious" to Feinberg and Mallatt:

> The reason why an ant colony can create "intelligent" emergent features
> the way a conscious brain does, yet not be conscious, is obvious.
> It is because the individual nervous systems
> within a colony of ants are not collectively woven together
> into a single functional and structural unity. (197)

Intelligence can be distributed variously between organisms and superorganisms—
and it makes a great deal of difference how—but in any case, it emerges from the way
they are wired together. To distinguish starkly between wired connections
inside a brain/body and wireless transindividual ones is to make a category mistake,
favoring chemical/electrical signalling that goes on in a brain over the pheromonal,
visual, verbal, and other signaling that goes on between creatures. The mistake
is not just in ignoring that they evolved in tandem; it is to make some of them
seem structural/hardwired and others "merely behavioral,"
some "physical" and others "occult" or even "supernatural."

Modern science emerged by distinguishing itself from religion,
which it characterizes as postulating occult entities,
but here you see how much the occulting is *its own*,
by favoring in-body connections to those between bodies,
connections within a species to those between it
and its fellow creatures and environments—all
reasons why ecology tends to be considered
a second-order science instead of the mother-of-all-sciences.
If you think the proposition that humans talk with trees
and that trees talk to each other sounds either druidical
in a primitive-religion or new-age way, *you're a modernist*.
Modern rationalism invented nature and the supernatural
by positing and entrenching the opposition between them.
The opposition is *something to think with*, a generative mistake,
a kind of motor (one that also spews out toxic by-products);
but actual *belief* in it keeps your thinking especially stuck.

Like Feinberg and Mallatt, evolutionist and famous atheist Daniel Dennett
makes a too-stark distinction between consciousness and everything else:

> Contrast a termite castle with Antoni Gaudí's wonderful church
> in Barcelona, *La Sagrada Familia*. They look similiar, but Gaudí's church

is a product of intelligent design: it's top-down, with a charismatic boss
who thought it out in advance. There is no Gaudí in the termite castle.

Excuse my tone, but Professor Dennett, have you ever designed anything
or ever thought or desired anything? Do you really believe that you are
the singular, top-down origin of your thoughts and designs and desires?

Unlike Feinberg and Mallatt, Dennett is at least willing to admit that "what we have
between our ears is more like a termite colony than you might be happy thinking"—
though as I hope you can tell, we *can* be quite happy thinking it! So it seems
there is no Gaudí in Gaudí either. This generates a puzzle: without a God—
that is, without a top-down "Intelligent Designer"—how did there come to be
an intelligent designer in the form of human consciousness? There didn't!

Dennett's problem is not that he rejects God *but that he sneaks him back in*,
in the form of human consciousness as a top-down intelligent designer.

Dennett's question (a version of what Feinberg and Mallatt's book addresses)
is "How on earth do you organize 86 billion neurons into Gaudí's mind?"

But Gaudí had already given him the answer, if he'd been listening:
La Sagrada Familia! The Holy Family! You are a sacred plurality
and a member of a sacred plurality! The holiness or sacredness—
if you like those terms, which I do—belongs to the teamlike,
three-or-four-member system-as-deity I sometimes call *M'sip*,
meaning "more than the sum of its parts," a.k.a. Figure 3.5B.

d   Qualia

Consciousness is often defined by tying it categorically to the emergence of *qualia*,
often defined as the quality of "what it is like": it is like something *to be me*
that no one else can know (whereas, by contrast, it is not "like" anything to be a rock),
and in my experience, an apple's redness differs from a trumpet's sound, as from yours.

To explore what's wrong with this, let's start again
with a humble example: the state of *being tattooed*.

I noticed, while being tattooed—and I've heard the same from others—
that, almost like adjusting a dial, I could interpret the ongoing sensation not as pain,
but as a more ambiguous signal. From the beginning, I was able to override
the response of *"pain: pull away!"* (in any case, I didn't jump back from the needle),

but as it went on, the vertical *overriding* gave way to more horizontal *displacement*.
It wasn't just that my reptile brain was telling me to pull away and being overridden
by my "higher" consciousness but that two competing sets of interpretations/emotions
were vying for control, with "This is interesting" coming to be the default mode.
No surprise, though, that when the pain surpassed a certain threshold, it became clear
(how shall I put this?) that *this was a dynamic process with a shifting balance*—
or in other words, *OUCH!* The tattooist could tell when I was tensing against the pain,
and she would say, "I can see that's a little *spicy*." She had found the perfect word
for the pleasure/pain border, and her syntax perfectly honored the ambiguity of referent:
what was spicy? Was it that area of my skin, that particular needle/nerve interaction,
or even that dynamical moment of our interpersonal interaction? My being on the verge
of pulling away (of shifting from willfully being an object to willy-nilly being needled
into being a subject) necessarily *triggered her* to prepare to pull back if necessary—but
is this different from saying that, in that moment, *she was moved to empathize* with me?

As this suggests, there's something not *higher* so much as *broader*
than an absolute "what it is like": It's like this, *and it isn't*.

I'm not a masochist, but I went far enough in the "This sensation is interesting" direction
to be able at least to imagine liking being on that edge. Should I say that I *discovered*
I'm no masochist, or I *decided* I wasn't (or to finesse it, that I was *inclined* to decide that)?
We can say at least that at that moment I was on the leading edge of "what it is like" and
"what I am like" and "what I like"—regardless of which direction I went with them.

If I started suffering from chronic pain, I would hope to be able to go much further
in the "This is interesting" direction. In fact, when I had a slow-to-heal knee injury,
I found myself feeling fairly constant, low-level irritability and anger.
I asked my doctor how it was that people in chronic pain were not just *assholes*,
and he laughed and said it was a phase that would change if the pain continued.
This was a piece of wisdom I needed to get from him since I was otherwise
so inside my own feeling of "what it's like to have constant pain" (or rather,
as I found out, what it is like *to be in the first few months* of that condition).
I downloaded his perspective—which was the perspective of many more people,
over more time, that *he* had downloaded—to leverage and displace my pain.

It's easy to guess why he didn't also mention
that some people in chronic pain *are* assholes.

In fact, as Feinberg and Mallatt point out, the capacity for prolonged suffering
is *an evolved one*. It requires more kinds of receptors and a broader range of feelings;

greater sensitivity and complexity. Bony fish (for example) only seem to be able to feel *acute* pain and respond accordingly; the ability to feel prolonged suffering is adaptive only for animals who benefit most from being driven to hide from predators as they heal. But it is conspicuously not the "highest" form of consciousness: a primary focus of much meditation, religion and related spiritual practice is on how to leverage prolonged suffering *not merely to transcend it*, not to be able to step outside it, but to transform it and oneself in the process in a way that makes one less of an asshole and, by the same token, also produces more empathy with assholes in chronic pain. Acquiring the ability to feel chronic pain may have been part of what makes us us, but this *us* is an in-between state, a necessary cusp of acquiring the ability to reprocess.

A system does not grow another dimension unless it's pushed, whether from the outside or by its own internal instabilities and contradictions (or typically, some mix of both). This is also how Wittgenstein describes philosophy: you have to have a problem and it has to bug you, deeply and sustainably bug you—a lot like chronic pain— if you are to be a philosopher. Some old philosophers suffer from "loss of problems"— which might itself be an interesting problem if you retain the capacity to be interested. But if you're comfortable, there's no reason to theorize or philosophize.

To put it another way, there is an effort involved in thinking in a philosophical manner, an exertion of consciousness, as if I were trying to jump out of myself—and being myself what amounts to a *burning building* is a highly motivating condition. These are the metagymnastics of self-consciousness that Feinberg and Mallatt describe (more on this below) as a mostly futile attempt to observe one's own neurons firing.

I exert myself and then fall back. My brain does something like what a gymnast does, landing from an aerial move, planting, recovering, running, coiling like a spring and jumping and twisting again (thinking can be convulsive!) and after a bit of this, I'm tired and turn to something else. But in off times, too, things can fall into place.

Let's say that, amid these exertions, I fail entirely to observe my neuronal patterns or that I only catch a several-steps-removed glimpse of my own operating principles before falling back into their orbit again. Yet in the exertion, in the failure and the falling back, *the operating principles and the orbit change*. This is evolution and learning, whether or not I learn to rest in a "higher" state.

Consciousness is the pearl we secrete around chronic pain
and we are the pearls being secreted around consciousness;
a set of origamilike folds in the topology of meaning.

What pain is it that our consciousness is being secreted around?
My own inclination is not toward the emptiness at the core of being,
the ultimately tragic and unfillable lack at the center of our psyches,
the giant black hole around which all our galaxies may revolve
(though these all are viable paradigms, as far as they go).
Get too close to *that,* and I am no longer *I* at all; otherwise,
We've been going in the "This is interesting" direction all along.

e   Self-reference
Reference is necessarily entangled with self-reference.

"Survival," in the Feinberg and Mallat account, "depends on neuronal networks referencing the outer world and the body, where the dangers arise against which consciousness helps to protect" (222). We're back to our first caveat: the story of the plucky little system versus the odds-stacked-against-it environment. The problem is in assuming an outer world (one that can simply be referenced, like *Law and Order*'s *offenders*). Just as earthquakes tended not to be very dangerous until we started building buildings and freeways and interconnected infrastructures that were vulnerable to them, there are no dangers except to a system that manufactures itself and them in the process. Jellyfish don't worry about Republican presidencies (though admittedly, this could be because they have no brains or because they stand to gain from global warming) and they don't suffer from suicidal depression; *we* made these problems by going out on the complex and precarious limb of self-consciousness.

According to Feinberg and Mallatt, neuronal "networks should not waste effort in consciously perceiving the detailed firings of their own neurons. That would be wasteful because these neurons are already protected by other homeostatic mechanisms (for example, by physiological mechanisms that maintain the constancy of the bodily fluids that bathe and sustain the neurons)" (222).

Accepting this, *for now*, consider that this is why, when the homeostasis of the system is challenged—as by a hallucinogenic drug or a migraine—we sometimes *can* see a manifestation of our own neuronal firings in the transculturally universal hallucination of complex fractal patterns that otherwise remain in the background of our thoughts and perceptions. Similarly, when your vision is tickled by the right kind of optical illusion—such as a printed starlike pattern that seems to scintillate—the movement you see is your own alpha brainwaves firing nine times a second.

Notice that, in the moment of seeing it, you do not naively believe
the image on the page to be flickering. Although it is surprising
to learn that you are seeing our own alpha wave frequency,
what you see *was* triggered by the image, which somehow interfaced
with your visual system like a key and lock; you see not merely
an inside-to-outside projection, but *a relationship* between them.

You might think that hallucinations and optical illusions and such
are unusual cases, but homeostatic disturbances of neuronal patterns
happen continuously and include *thoughts, feelings*, and *sensations.*

Feinberg and Mallatt's principle of "auto-ontological irreducibility"—the idea
that "the subject cannot experience the workings of his or her own neurons"
and that, accordingly, "neural processes that create sensory consciousness refer
all feeling states away from the brain itself to something else" (221) *is itself
the self-reflexive characterization of the attributive style of a scientist.*

Dangers aren't just outside the mind (in the world or the body)
and same with the pleasures; systems have internal contradictions
that drive them forward as well as arcs of arousal, frustration and
resolution and release that they are compelled to recreate and to ride.

We are far-from-equilibrium systems, riven and driven by incompleteness,
restless contradiction and precarious instabilities. We are wobbling tops,
guttering flames, burning buildings, turbulent rapids.

You may find this too melodramic, the rhetoric too purple, too poetic,
but as such it is a corrective to the suburban dream of homeostasis
suggested by Feinberg and Mallatt's account of neurons bathing
in their sustaining fluids—reading the *New Yorker*, no doubt,
with martinis in hand and Mozart playing softly in the background.

13   Conclusion

> At the time the earliest vertebrates were constructing
> their first mental maps of the external world
> from their exteroceptive sensors, they were also
> imbuing their interoceptive and exteroceptive sensations
> with affect, value, and therefore full sentience. (168–69)

It is a beautiful insight, situating the emergence of consciousness—"full sentience"—
at about 350 million years ago in the first fishlike vertebrates, and it goes a long way
toward undermining the cognitivist and exceptionalist accounts of consciousness
as belonging to humans alone. But as we've seen, there are a few problems.
The sentence assumes agency and subjectivity, assumes the outsideness of
the world to be mapped and the insideness of the mapping. Things look different
when we start instead from the premise that orienting oneself in the world—
negotiating any engaged relationship—involves *continuous* mapping.
Even the hypothetical, ambiguous amoebalike creature with which I began,
which swam up a chemical gradient toward food, was doing the same thing.

The apparent unity of an apparently external referent—
qualifying as a marker of consciousness for Feinberg and Mallatt—
can be created from the coprocessing of signals from different senses,
as in the ambivalent situation when my nose is telling me *very interesting*
but my eyes say *run away*. This unity is not a given that consciousness
has to evolve to match or grasp but another "intelligent emergent feature"
of a brain that cognitive scientist Marvin Minsky described as a "society of mind"
and of creatures and ecologies and maybe living planets—trans-systems that think—
like Steve Shaviro's paradigmatic "many-headed" slime molds.

If you are inclined to resolve this ambivalence by attributing it to a unified object (if this
is your attributive style), then you have a very challenging psychodynamical problem.
You can do years of therapy to overcome it, or if not, then (like an external predator)
it will keep coming back and biting you in the ass. Your therapist might tell you:
those are *your feelings*, not unproblematically features of the other person,
and they're valid and it's okay to have them, but your feelings aren't just
about me (for example) or about you, but about how you and I and we
co-construct each other, in the therapeutic relationship, *as in love*,
and also sometimes *as in reading*, which is why loving someone,
even reading someone, especially for a long time, can be how
we replay and project our fears and fantasies
and, by the same token,
can be healing.

CHAPTER 7

Blake Magic

1 Famous in Heaven

From Lincoln's magic militarism to Shelley's visionary nonviolent resistance—which, after a weak start (a complete failure, actually) came to contribute to a number of real political struggles—we can go a step or two further.

During his lifetime (1757–1827), William Blake was a marginal artist and writer whose major works were seen by maybe a hundred people.

Though he was capable of disappointment and even despair, mostly he continued to assert that, while a virtual unknown—at least, on earth—he was, as he put it, "famed in heaven."

He and his wife, Catherine, survived on his illustration work (he worked as an engraver) and on the occasional purchase, by one of his few patrons, of his own hand-etched, colored, and hand-printed books and other works.

All but one of Blake's books are what we would now call "self-published" in tiny editions. By writing, illustrating, engraving, printing, and selling them himself, he had cut out all the middlemen—and most of the possibility of gaining an audience.

In a series of illuminated manuscripts, Blake invents a kind of *religion*, complete with its own philosophy, mythology, deities, ethics, and aesthetics. If a religion had emerged from 1960s counterculture, Blake might have found a place in it.

Musical records of the rediscovery of Blake by beatniks and hippies include his poems sung by Allen Ginsberg, the Fugs garage-band rendition of his "Ah, Sunflower" and their country-inflected version of one of his first poems, "How Sweet I Roamed."

A snippet of Blake's poem "Auguries of Innocence" is crucial to the song "End of the Night" by the Doors, who took their name from Aldous Huxley's 1954 account of tripping, *The Doors of Perception*.

Huxley had taken his title from Blake's *Marriage of Heaven and Hell*: "if the doors of perception were cleansed/every thing would appear to humans as it is, Infinite." (and yes, by the way, I changed "man" to "humans." Blake told me to, so it's okay.)

Later, singer Van Morrison and punk pioneer Patti Smith became genuine Blakeans. Smith's "my Blakean year" (2004) tells how she found her own path as an artist, and she went on to publish a book of poems under Blake's title "Auguries of Innocence."

As it happened—and in case you hadn't heard—it was not religion but *literary criticism* that laid claim to Blake's legacy. Too bad.

Literary criticism was invented to claim aesthetic value for particular kinds of texts, but it tends to neutralize any other value they might have. The field is still dominated by *historicism*, which focuses on where a text comes from and the forces that shaped it.

As I have already said, my focus is not on the past but on texts as *resources for possible futures*. Accordingly, I read Blake's works as *texts that might still be held as sacred by some future religion*.

I don't believe this will come to pass, but whether or not it could happen is entirely beside the point. What the text does in the world, *its meaning*, is what happens on the way, even if it always fails to become a religion.

If it somehow did become a religion, I would probably have to renounce it!

> ASIDE: No Success Like Failure
>
> (a) One of my mentors, Michael Sprinker, was a true believer in the core principles of Althusserian scientific Marxism.
>
> He used to say wryly that, over his career of teaching and writing, he hadn't managed to convince a single person of its validity. (Well, there was one guy at Oregon State, but the basic point still stands.)
>
> Alongside this apparent record of complete and lifelong failure is the *transformative influence* this passionate, generous scholar had on nearly everyone (including me) who came into his orbit.
>
> (b) Chilean filmmaker Alejandro Jodorowski's *Dune* has been called "the most important film that was never made." Jodorowski assembled

teams of artists who sketched out visionary concepts, sets, and scenes.
Many of these, after the studio killed the film, found their way
into some of the most important sci-fi films of the late 20th century.

In his script, Jodorowski changed the ending of the novel: in his version,
the main character is killed but the others begin to speak with his voice.
This turned out to be prophetic: the film was killed by the studios
but ended up speaking through films that took up its failed promise.

2 Blake Magic

It's easy to say Blake's works were fated to find an audience,
but as they also say, hindsight is 20/20. At his death in 1827,
each of his major works existed only in handfuls of copies.

Some had even more precarious footholds in this world.
There existed only *a single, handwritten copy* of ten poems
that came to be known as the "Pickering Manuscript."

These poems survived by the slenderest thread.
They were literally *marginal*: Blake wrote them out
on paper trimmed from the margins of a failed project.

William Hayley, for a while Blake's chief employer, had written ballads
based on anecdotes about animals. His idea was for Blake to print for sale,
by subscription, one ballad per month, accompanied by Blake's engravings.

Hayley described these as "vehicles contrived to exhibit the talents
of my friend for original designs and delicate engraving." In fact,
it's hard to imagine a project better contrived for Blake to *fail at*.

Blake did his best to lend sublimity to Hayley's verses, to tone down
his own visionary style and adapt it for a genteel audience: the result
is an awkward mix of kitschy realism and stiffly posed mythic figures.

When it came to selling out, try as he might, *Blake sucked at it*.
"Sales were less than brisk," so after four ballads, they gave it up.
Blake cut off the wide margins of the unsold sheets to draw on.

He also wrote out poems on the cut-off margins and sewed the pages
into a little pamphlet. Eventually—in 1865, 38 years after his death—
antiquarian Basil Pickering bought the pamplet for 7 pounds 5 shillings.

Included is "Auguries of Innocence," the poem that would go on to be taken up
by Jim Morrison and Patti Smith. It's a collection of two- and four-line aphorisms,
beginning with what would become Blake's most famous mystical formula:

> To see a World in a Grain of Sand
> And a Heaven in a Wild Flower
> Hold Infinity in the palm of your hand
> And Eternity in an hour.

Just as a two-dimensional plane holds an infinitely long line and three-dimensional space
holds an infinitely wide plane, we could start by saying that an even higher dimension
holds the time and space that stretch out in all directions. Call this dimension *meaning*.

Meaning inheres in the fractal nature of the universe;
in the way that macrocosms and microcosms (meta-
and subworlds; systems, large and small) are linked.

The poem is dedicated to showing how this linkage differs—*radically*—
from the hierarchy-of-hierarchies known as the *Great Chain of Being*,
a dominant paradigm of universal order starting in the Renaissance.

In the Great Chain of Being, a macrocosm ruled by God the Father is echoed
(as in a series of nested dolls) in earthly kingdoms ruled by Kings, families ruled
by patriarchal heads-of-household, individuals ruled by Reason. *Blake hated this.*

In the (Blakean, and later, postmodern) alternative, as Bruno Latour puts it,
"scale is the result of the *number of connections* between localities, not
the circulation through any preordained zoom from very big to very small."

A human being or a collective ecological entity (such as our planet) is not built
by adding hierarchically larger and larger layers. Systems and their components
only emerge together, small and large, along with the connections among them.

The whole remains part of the parts. If you get this ecological understanding
of scale, with the sense of the interdependence of the large and the small,
system and components, you can wiggle out from the Great Chain of Being!

We have to revise the notion of higher dimensions, which may be curled
up in lower. Something of greater complexity may be the sub-system
of something lesser. To hold doesn't mean to contain. Systems emerge
from other systems not by transcending them but as connected networks.

Blake's *Grain of Sand* is tiny and crystalline; it is among the lowliest things
with enough emergent order and complexity to be considered a proto-entity
(as DNA, on its route to being discovered, was called an "aperiodic crystal").

Blake's *Wild Flower* is both ephemeral and unique (it's wild because
it can't reliably be reproduced) and thus is another kind of example
of the lowliest thing that might still qualify as a "crown of creation."

The sandgrain and wildflower are equivalent or greater in complexity
than the worlds in which they participate, just as nonlinear systems
can dimensionally hold and enfold linear infinities and eternities.

But I am not going to be making a literary-critical claim about the poem.
Though a certain amount of interpretation may be necessary *to get there*,
my basic claim is not interpretive at all but *factual* and *performative*.

This claim is the bare fact of the poem's survival
through its tenuous early existence, the ongoing

dance of its coded letter-strings with living webs
of neurons that happens whenever it is read.

If you are reading this—and at least the few bits of the poem quoted here—
that fact's established: not by Blake's or this text alone but *by you* as you read,
in the linkage between the different moments you and I each call *this moment*.

This is also the *performative* claim: this essay, along with your reading of it,
is the survival of the text. It is not the fulfillment of a heroic aspiration—
survival generally isn't—but fact by virtue of its own stubborn persistence.

This wouldn't mean much: lots of small and insignificant things
survive in out-of-the-way places. Some are inevitably rediscovered,
found to have great value and significance, and sold on the internet.

But this text is not only a small thing that has survived: it also happens to be
about the survival and significance of small things. That's it; that's the heart
of the magic or the magic trick. (I have *even smaller claims*, if you're interested!)

A dodgy proposition when it was written, the poem came to *enact*
what it's about, by being literally and figuratively marginal but surviving
and gaining a foothold, linking its referential and performative meanings.

In spite of its long string of hyperbolic assertions, the performative thrust of the poem
is not the making of any argument. As with the Shape in Shelley's "Mask of Anarchy,"
it operates as a Romantic Symbol, part of what it represents. The poem proclaims
its kinship with the marginal, the grain of sand and the wildflower. It wagers itself.
The extent to which it survives is the extent to which it will have been proven.

3 Auguries

Auguries are divinations of the future by reading various signs and omens.
In ancient Rome, these omens included the flights of birds and other animal behaviors.
Bird omens were known as *auspices*; their official interpreter was known as an *auspex*.

For Blake, a systemic kind of ecology binds the small and large, the local and global,
so that a small sign—a distant rumbling sound (think of Asimov's "Sound of Thunder"),
a slight tremor—might well be the token of something apocalyptic (good or ill) to come.

Blake's longer prophetic poems often involve attending to these "canaries in the coalmine," as does another one of his most famous poems, "London":

> I wander thro' each charter'd street,
> Near where the charter'd Thames does flow.
> And mark in every face I meet
> Marks of weakness, marks of woe.

What the wandering prophet sees is how the sighs and cries of those used up by war, religion, and patriarchy (the latter via marriage and prostitution) return to haunt those institutions.

As he puts it in "Auguries of Innocence," what the prophet hears and sees is how "the Harlot's cry from street to street/shall weave old England's winding sheet" and how "a dog starvd at his Masters Gate/Predicts the ruin of the State."

The principle here, as explored in previous chapters, is that the sustainability of any collective entity depends on how it respects (or doesn't respect) its humblest components and interdependent systems and subsystems.

The most apparently expendable components or entities can have agency all out of proportion with their apparent insignificance, as is the case with what in ecology are known as *keystone species,* which might be weevils or wasps, lowlier even than Blake's birds and small animals. Entities depend on lower-order components and forget this at their peril.

Blake put the core principle succinctly in his letter to an editor (who, as you might have guessed, declined to publish it) on the occasion of the arrest and imprisonment of an astrologer:

> The Man who can Read the Stars often is opressed by their Influence,
> no less than the Newtonian who reads Not & cannot Read
> is opressed by his own Reasonings & Experiments.

The way we make our worlds as the producers and products of precariously far-from-equilibrium, nonlinear systems beset with dangerous contradictions is not an arcane insight of complex systems theory. It has long been common sense.

"As you sow, so shall you reap"; "what goes around, comes around"; "live by the sword, die by the sword";

consequences of your acts "come back and bite you in the ass";
or as per Marx, the "bourgeoisie produce their own gravediggers."

If you are focused on faith or belief, you might be inclined to say that these aphorisms reflect a rather naive confidence in some fundamental moral order in the universe—
or instead, a bad-faith whistling-through-the-graveyard of a clearly amoral universe.

But if you focus on practices and performativity instead, the aphorisms point
to *opportunities for nonlinear system-building* (cocreating self and world)
and to the consequences of failure to recognize these opportunities.

These are the places where magic, science, and religion coincide.

Martin Luther King's assertion that "the arc of the moral universe is long,
but it bends toward justice" (adapted from abolitionist Theodore Parker)
cannot be understood as a naive referential statement about a universal bent.

It opens up when you take it not as a descriptive statement but as *an act of holding and bending the arc*. Tension is sustained by grief over the arcs of so many lives bent and cut short by injustice, grief no consolation can hustle through its stages.

Practice and faith weave together *in practice*, but the focus on practice
doesn't exclude making referential observations or ontological claims
about the nature of the universe. In fact, it seems to require at least one.

It must be a universe with the capacity to form systems—
even in their most reductive form: self-fulfilling prophecies
that can, if they are successful, edge into real prophecies.

> ASIDE: Prophecies
>
> By the mechanism Freud called *the return of the repressed*,
> your actions can end up bringing about exactly what you fear,
> which may strike you as Fate, coming from the world at large.
>
> Think of Laius, sending his baby son Oedipus off to die
> in order to thwart the prophecy that his son would kill him,
> triggering, in the fullness of time, just what he sought to avoid.
>
> When he meets his adult son on the road, he fails
> to recognize him. They fight and Laius is killed.

> Freud's Oedipus Complex seems less definitive for masculinity
> than the *Laius Complex*: fathers paranoid about keeping power
> are aggressive and cruel to sons, who are cruel and agressive in turn.
>
> (We should give credit to that precocious teenager Mary Shelley
> for nailing this narrative—and to those who, by a brilliant mistake,
> apply the creator's name to his otherwise nameless monster.)
>
> The story illustrates what I proposed earlier: tragedy can be based
> not just on *hubris* (thinking one has *powers one does not have*)
> but on failure to recognize *powers one has* to co-create the world.
>
> Toxic masculinity and the Anthropocene in a nutshell:
> the consequence of failures to recognize opportunities
> for nonlinear system-building—failures of magic.

The universe rewards exploration of opportunities for system-building along the *via of viability*: developmental pathways that knit together creatures and the ecosystems they shape and are shaped by.

4   Angel Kings

In Blake's "The Grey Monk"—another poem in the Pickering Manuscript— a monk, who's been tortured and his family hounded because of his writings against the tyranny to which they are subject, makes a prophetic claim:

> But vain the Sword & vain the Bow
> They never can work Wars overthrow
> The Hermits Prayer & the Widows tear
> Alone can free the World from fear
>
> For a Tear is an Intellectual Thing
> And a Sigh is the Sword of an Angel King
> And the bitter groan of the Martyrs woe
> Is an Arrow from the Almighties Bow

The earlier stanza is easy enough to understand, in principle. Violence begets violence, so, as Audre Lorde famously put it, "the master's tools will never dismantle the master's house."

The means are *part of the ends*: we organize not simply as a means,
but because the kind of organizing we do *is* the end. So far so good.
But the next stanza seems more like magic thinking: our pains
will come back to haunt and defeat our oppressors. *Really?*

Through affective labor, feeling and desire (tears, sighs, and prayers)
bear at least the potential to *organize* and to *be organized* into dissent
via emergent, collective, potentially transformative *structures of feeling*.

Even the sighs and cries and prayers are not possible
without the *capacity to imagine things being otherwise*.
Blank despair doesn't cry. "Fear & Hope are—Vision."

By calling tears *intellectual*, Blake displaces any oppositions of physical, cognitive,
emotional, spiritual (the latter is closer to what *intellectual* meant in Blake's day).
He suggests a link with some power that authorizes them or is authorized by them.

This power is not only a potential. It must already exist, if only
in the imaginary. If you don't like the language of *organizing*
and *organicism*, call it *an abstract machine*: there is a circuit
of self-organization in which both system and components
are already being manufactured. The potential entity—
not just an *angel* but an *Angel King*—*already exists*.

This can be understood—for a start—in a nonmystical, historical sense.
Like many who opposed capitalism, just as it was solidifying its rule,
craftspeople (like Blake as an engraver) were among the most radical.

Craftspeople were empowered by a precapitalist tradition,
a medieval guild system that organized and valued their labor—
and accordingly, they had a lot to lose by being proletarianized.

Having completed an apprenticeship as an engraver
and achieved at least the rank of journeyman,
Blake came honestly to his radicalism.

And lest you think his radicalism was confined to mystical texts,
Blake was among the mob that stormed Newgate Prison in 1780,
released the prisoners, and burned down the prison. And later,
he was charged but acquitted of sedition for saying *Damn the King*.

Nobody speaks from pure empowerment or pure disempowerment.
If there were only one power hierarchy, subalterns could neither speak
nor exercise agency. Axes of empowerment and abjection cross in us all.

This is as true for the overprivileged bully driven by secret fear and shame
(patriarchal, heteronormative masculinity in a nutshell) as for the abject
sustained by secret reserves of strength, dignity, vision, or plain perversity.

At bottom (as I've mentioned), the fantasy of omnipotence
underlies our ability to feel and act and to have agency at all
because we have been both abject and dependent—*and loved*;

because we've been sponsored and authorized by an angel king—
usually in the person of a *phallic mother* (psychoanalytic term)
whom we have variously internalized and actualized ourselves
and *continue to constitute in our connection with others.*

Starting with *introjection* turns inside out the supposedly demystifying claim that God
is a *projection* of parental unconditional love, power to punish and reward, and so on.

In psychoanalysis, the phallic mother is usually cast as a regressive and defensive fantasy
that operates as a refusal to recognize "castration"—that is, to recognize that women don't
have penises and nobody has the "phallus" (a symbolic scepter of kingly omnipotence).

Alternately, this phallic female deity can stand for the magical and necessary mix
of abjection and omnipotence—for how we must be both creatures and co-creators
of our network of relationships and worlds to build and sustain living systems.

To take it in the direction of religion, say that this knitting—
the places or interfaces where the whole is in the process of becoming
more than the sum of its parts—are where the Angel Kings preside.

5   Magic in the Current Conjuncture

We experience multiple economies, at least multiple personal/affective ones,
as attested by pop songs: think of REM's "with love come strange currencies"
or the Beatles' "in the end, the love you take is equal to the love you make."

Is the personal and affective simply the realm to which the remnants
of an old "gift economy" have retreated as capitalism keeps extending
its hegemony ever further into our hearts and minds? Yeah, probably.

Still, if you can't believe we're entering the eternal, seamless reign of capitalism
at the "end of history"—or that contradictions can be exploded once and for all
to usher in a lasting dictatorship of the proletariat—where does that leave us?

The longer answer involves exploring how the contradictions of capitalism
can be activated in the direction of life and evolution—not necessarily all at once,
but in everyday survival. Blake's "Auguries" have something to say about this:

> The Princes Robes & Beggars Rags
> Are Toadstools on the Misers Bags

Accumulations of capital (the miser's money bags) produce extremes
of wealth and poverty, power and abjection. Parasitism and poisons
(embodied in the toadstools) are bred by stagnation (congealed capital).

This was not an unusual view in Blake's time, when the logic
of capital had not yet saturated so many dimensions of human life.
Precapitalist common sense runs deep even in words themselves.

This is why *the condition of being a miser* is called *misery*—same reason
privation and *deprivation* are rooted in *privacy*: both involve being cut off
from dynamic give-and-take in the networks that constitute social being.

And this idea—that money should flow in social use,
not be hoarded in capital—is still not so unusual today.

The clearest (even if the most reductive) accomplishment of the Occupy movement
may well have been naming "the One Percent" as a parasitical coagulation of capital—
and the movement's *affective labor to situate the rest of us toward and against it.*

> The Strongest Poison ever known
> Came from Caesars Laurel Crown

Consolidation of wealth/power—fostering the notion that there is only *one* arbiter
and *one* hierarchy of value—may begin as economic but proceeds to infect other realms.

"One command, one joy, one desire,/One curse, one weight, one measure/One King,
one God, one Law," Blake's Urizen thundered in his disastrous bid for godly supremacy.
Note Blake's elegantly and radically negative recoding of kingship *and monotheism*.

The laurel crown was awarded not only to military victors but to poets
who won sanctioned contests. Again, Blake isn't spinning arcane doctrine
but restating the populist notion that society "rots from the head down."

Monarchy infects everything, or as Tom Paine put it, "hereditary despotism,
resident in the person of the king, divides and subdivides itself
into a thousand shapes and forms, till the whole of it is acted by deputation."

Even if all evils could be said to derive from kingship—just as money
is said to be the root of all evil—this also explains why revolution
could never succeed simply by cutting off the king's head.

Fortunately, other kinds of power and value continue to exist. As Blake puts it,
"One Mite wrung from the Labrers hands/Shall buy & sell the Misers Lands."

A different kind of value, based on labor, not only exists alongside capital, but
(is it even imaginable?) will come to take priority over capital—or, even now,
takes precedence in some of the simultaneous realms in which we live.

This is related to Walter Benjamin's assertion that, though "spiritual things"
cannot exist without the material basis on which they can be developed,
spirit manifests itself in class struggle as "courage, humor, cunning, fortitude."

These have "retroactive force and will constantly call into question every victory,
past and present, of the rulers." You could say (mystically) that this force
comes via empowering emanations that travel to us backward from a future

in which we will have prevailed. Or you could say, even in the darkest hour, we bear this future within and among us—an absent presence, an *Angel King*.

Is this the best we can hope for, then—not that we will win, but that our persevering on the margins will "constantly call into question every victory, past and present, of the rulers"?

Taking anarchism seriously and rejecting too-singular consolidations of power and value, it seems that we must also reject the notion of a final reckoning between forms of value, a singular meta-economy that trumps them all.

Even so, we are far along in what political theorist Wendy Brown calls "Neoliberalism's Stealth Revolution": the takeover of ever more realms of human experience by an economic logic of investment and return.

Taking Brown's example from my own line of work, the worth of an education is increasingly understood as simply the amount that it will add to your income.

The notion that education *creates nonmonetary forms of value*— the way it used to be said that it fosters the "informed citizens" necessary for a democracy, or enables people to live the "good life" by becoming "cultured"—now seems naive or incomprehensible.

The cultured and informed citizen looks now like the shabby Victorian aristocrat with a title and a castle but not much liquidity—survivals from previous systems who must either figure out how to monetize their assets or quietly fade away.

I'm not saying the *ancien régime* should have been bolstered indefinitely: the notions of college making you "cultured" and an "informed citizen" was mostly a threadbare code for reproducing privilege. *Mostly*.

When funding for the State University of New York (where I used to teach) was being cut in the 1990s, our provost wrote an op-ed piece in which he claimed the university is a *cash machine* (his words) for the local and regional economy.

Not having come to grips with being a cog in a cash machine, I was horrified. But I also knew it was the "correct" argument to make—that it was strategic to remind people that investing in public education brings economic returns.

That was 20 years ago. The "Stealth Revolution" has continued to metastasize, more turnings of the screw in the ascendance of exchange value over use value,

more grief for those squeezed by it. As Marx put it in 1848, the bourgeoisie
"has left remaining no other nexus between man and man than naked self-interest."
In other ways, though, quantitative change seems to be becoming qualitative.

This is the case with the rise of the *attentional economy*, in which we "add value"
to whatever we look at and click on—and can be valued by the same process,
which colonizes the personal via the logic of Likes, Friends, and Followers.

The idea that you might have some *identity* or *integrity* distinct
from what otherwise might be called *branding*—something
that would be debased or jeopardized by a price tag—is so *passé*!

This is not exactly news. In Wilde's 1895 *Importance of Being Earnest*,
when Algernon asks Cecily if he might look at her diary, she rebuffs him
in a way one could easily update to make her a sassy blogger:

> Oh, no. You see, it is simply a very young girl's record
> of her own thoughts and impressions, and consequently
> meant for publication. When it appears in volume form
> I hope you will order a copy.

Wilde parodies ideology by saying aloud its *open secrets*.
As Marx put it—in the voice of prophecy, not parody—
the bourgeoisie drown religion, manners, feelings
"in the icy water of egotistical calculation."

Finally, to pick an example not so easy to push away with moral condemnation,
listen to Hugh Herr's account of how far we've come (thanks in part to technologies
developed by him) in the use of bionic technology for people who have lost limbs:

> "For the first time in history we've normalised walking speed and its energy cost.
> In other words, if you simply measure a user's speed and energy expenditure,
> you can't tell whether they have bionic legs or biological legs."

Note how naturally the issue comes down to a logic of cost and expenditure—
an update of what happened in Blake's time when the new notion of "horsepower"
enabled the measure and monetizing of steam technologies, now *closer to home*.

This goes beyond recognizing the *cash machine* argument as *correct*. Here,
no argument needs to be made that the cost/expenditure model is the right way
of thinking about this (or any) particular issue. It has become fully *epistemic*.

An *episteme*—a broader and deeper kind of *paradigm*—is not an argument.
It *sets the terms by which arguments are made*—a form of universal currency,
the way a banknote claims to be "legal tender for all debts public and private."

It goes beyond ways of talking or thinking (as the term *paradigm* is often used),
just as making things interchangeable via money *changes the things*. It is,
like other phenomena this book explores, a way of making and remaking worlds.

You can deplore young narcissists, but could you be against artificial limbs?
If expenditure and investment/return really is or has become the key
to everything from education to artificial limbs, what's left?

If an episteme is, by definition, something you can't argue with, something
you can't step outside—if it defines what counts as rationality—what's left?
If you don't know, I suggest you read this book again, getting at least as far
as *the first word of the title*.

To return to the example of my own line of work: education
is filled with people—teachers and students—whose motivations
are not intelligible by the model of investment and return.

What advantage do those who understand this have? Does this knowledge
better enable you parasitically to exploit the energies, hopes, and dreams
of students and teachers; to better take more of their money and their labor?

Or can it be leveraged to pry open livable spaces and networks
in an otherwise suffocating and increasingly unsustainable system?

In the end, Blake's "Grey Monk" swings from what might seem like
magic thinking about *the potential of the meek to inherit the earth*
to what might be a depressive story of *a never-ending cycle of violence*.

> The hand of Vengeance found the Bed
> To which the Purple Tyrant fled
> The iron hand crushd the Tyrants head
> And became a Tyrant in his stead.

So which is it? The shortest answer is: *you pick*.
The first political task—the magic, the delusion—
is the feeling that *another future is possible*.

6   **Five Asides**

ASIDE: Black Lives Matter and Judith Butler

Activists Alicia Garza, Patrisse Cullors, and Opal Tometi—
queer women of color—named and catalyzed *Black Lives Matter*
after white vigilante George Zimmerman was acquitted
of murdering unarmed black teenager Trayvon Martin in 2013.

Cullors picked up the phrase *Black Lives Matter*
from an email message sent by Garza and used it as a hashtag.
Three simple words form a short sentence, a manifesto in a meme.

A name doesn't make a movement, but this one does some heavy lifting
as a speech act. It manages to *speak from* empowered personhood,
from the unequivocal existential and emotional reality of the speaker.

At the same time, it *speaks to* a political reality in which it remains
a radical proposition—to a system committed to denying it in fact
and denying its own denial. It *activates* this contradiction.

The "colorblind" version *all lives matter,* far from universalizing,
practices precisely this double negation and erasure.
Philosopher Judith Butler elaborated this point:

> To make that into a living formulation, one that truly extends
> to all people, we have to foreground those lives that are not
> mattering now, to mark that exclusion, and militate against it.

In fact, Butler's previous philosophical work resonates deeply
with the *Black Lives Matter* movement. Her book *Bodies That Matter*
(1993) studies bodies as sites of both political subjection and agency.

Her *Precarious Life: The Powers of Mourning and Violence*
(2004) focuses on differential values of lives and the question
of which losses are allowed to be publicly mourned.

Whatever Butler's work may have done to enhance the resonance
of these ideas prior to Cullors picking out Garza's phrase as a hashtag
and the hashtag catching fire, this is not an "influence" argument.

It is an argument for recognizing the common affective labor—
the deepening and broadening and politicizing of the resonances—
being done by Garza and Cullors and Butler and many others.

The resonances testify to common experiences of precariousness
and subjection to violence—however differentially (psychologically,
racially) we come *from them* and must keep coming to terms *with them.*

Butler had long been faulted both for the difficulty of her prose—variously
described by haters as obscure, incomprehensible, vague—
and for the notion that such philosophy could have political effect.

Political philosopher Martha Nussbaum went to an extreme,
calling Butler "the Professor of Parody." By indulging
in the navel-gazing of so-called "identity politics"
(and, Nussbaum implies, by focusing on queer identities),
Butler pulls women away from "real" feminist politics,
and thus her "hip quietism ... *collaborates with evil.*"

These old slurs hardly need countering (as Blake said, "Listen
to the fools reproach! it is a kingly title!"). The demonizations
and other repercussions are *part of the impact of Butler's work.*

Even so, the counterpoint—an expanded notion of political action
that includes affective labor, and performative (even magically
circuitous) effectivity—is part of my main argument here.

It is affiliated with the argument that many others,
including Garza and Cullors, make for the vitality and necessity
of queer theory and activism for other kinds of political action.

It is an affirmation of the power of the work of grieving—
the Widows tear—and of the affective and political labor
of intellectual work—*the Hermits Prayer*:

> But vain the Sword & vain the Bow
> They never can work Wars overthrow
> The Hermits Prayer & the Widows tear
> Alone can free the World from fear

> For a Tear is an Intellectual Thing
> And a Sigh is the Sword of an Angel King
> And the bitter groan of the Martyrs woe
> Is an Arrow from the Almighties Bow

I once saw Butler speak at CUNY Graduate Center. She stood
at the podium, voice crackling with intellectual energy and precision.
Facing her was a row of scowling, white-bearded senior professors.

I flashed on a scene from the New Testament that I didn't even know
I knew (from assorted paintings), in which the twelve-year-old Jesus
is accidentally left behind in Jerusalem by his parents.

"They found him in the temple, sitting in the midst of the doctors,
both hearing them, and asking them questions," and when he spoke,
"they understood not the saying which he spake unto them."

Take that, haters! Every demonization deserves a deification,
every denunciation to be answered with a joy. If I thought reason
would suffice, I wouldn't be advocating Magic Science Religion.

Anyway, it was the young-versus-old part of the dynamic
that triggered the "Christ among the Doctors" image for me.

Intelligence does not triumph automatically over entrenched authority,
but sometimes it can figure out how to hold its own, and even
how to leverage terms by which it otherwise would be discounted.

This *performative intelligence* is part of the perverse work of the child,
which is why the "Christ among the Doctors" dynamic is so fitting.

I found myself thinking of the everyday perverse magic of survival
as performed by transgender kids in particular, and I flashed
on the following lines from Blake's "Auguries of Innocence":

> He who mocks the Infants Faith
> Shall be mockd in Age & Death
> He who shall teach the Child to Doubt
> The rotting Grave shall neer get out
> He who respects the Infants faith
> Triumphs over Hell & Death

What kind of speech acts are these? Curses? Romanticized
assertions of childhood wisdom? Simple referential truths
about how what goes around comes around? All three?

This is where I want to leave it. Having recruited Blake
for Black Lives Matter, queer ecofeminism, systems theory—
and the Magic Science Religion I've been advocating—

my work (for the moment) is done. I don't want to tie it with a bow,
but, here at the end, to let go of what my training taught me to do:
to explicate, anticipate objections, and so on.

> A Riddle or the Crickets Cry
> Is to Doubt a fit Reply

Blake's works are performative questions, riddles, parables, and koans—
more so than words in their referential dimensions. The opposite of doubt
isn't faith but the practice of performative magic—and dogged persistence.

The cricket's cry is fitting because it has the prelinguistic,
rhythmic music of poetry, and because it is such a tinny
and tiny sound that establishes itself by such persistence.

Its efficacy is not in itself but as it is tuned or orchestrated
into a *collective* acoustic entity. *Though I may be alone,
I feel you all around me, now and in the past and future.*

And here, for the moment, my own chirping subsides.

ASIDE: An Elegant Pose

In Borges's 1940 story "Tlön, Uqbar, Orbis Tertius," a secret society
works for generations to produce an encyclopedia of an invented planet
called Tlön, and after it's found, the world begins to fall under its spell.

> The contact and the habit of Tlön have disintegrated
> this world. Enchanted by its rigor, humanity forgets
> that it is a rigor of chess masters, not of angels.
>
> Already the teaching of its harmonious history
> (filled with moving episodes) has wiped out

the one which governed in my childhood;
already a fictitious past occupies in our memories
the place of another. The world will be Tlön.

I pay no attention and go on revising,
in the still days at the Adrogue hotel,
an uncertain Quevedian translation
(which I do not intend to publish)
of Browne's *Urn Burial*.

As everything morphs around him and all are enraptured
by a shiny but ultimately empty New World Order,
the narrator turns his back, content to be left behind
to antiquarian musings in the faded charm of his retreat.

Having been involved in the initial discovery of Tlön—
as Borges was an early explorer of literary postmodernism—
the narrator styles himself a melancholy and ironic Moses
dying on the shores of what will be a false promised land.

It's an elegant pose for an old person with enough privilege to strike—
and I find it rather attractive, though elegance isn't in my repertoire.
I see myself as more of a crank in a trailer home on the edge of a desert.

There is natural comradery among the old aristocrat, the craftsperson,
and the newly proletarianized. We know the future cannot belong to those
on the unsustainable peaks. The only way ahead from there is *down*.

How will the precapitalist past—preserved sometimes in feelings
and common sense and even in the etymologies of words—
join up with another future whose spark it will have helped keep alive?

Which eddying backwater stands a chance of becoming the main channel?
Which struggling mutants will come to define the course of evolution?
Meanwhile, "to work, even in poverty and obscurity, is worth while."

ASIDE: Ritual Retraction

I'm sorry to have written this, and I hereby retract and renounce it all.

Sorry to have bastardized the magic, the science, the religion—
exactly what I hoped to defuse upfront by announcing
that I would (consciously and strategically) be doing so.

Sorry to have been flip about things I should have engraved on stone
(and vice versa), sorry to have reduced to aphorisms and cliches
what I should have explained painstakingly step by step.

I'm sorry to have so compromised the content by the form. Sorry
to have been boring—and sorry to have tried too hard not to be boring.
Sorry for both my own sloppy laziness and my mechanical work ethic.

I'm sorry that the perverse and combative contrarian in me
is so often overruled by the touchy-feely and the sublime,
and I would also be very sorry if this were not the case.

I'm sorry to strike *this* pose, trying (I surmise) to defuse criticism,
to get credit for humility, to backhandedly admit as weaknesses
what I hope are strengths or part of a package deal *with* strengths.

I suppose the reason it feels so good to say all this is that,
while *referentially* it complies with various forces of negation
(making me feel I'm flying under the radar), *performatively*
It's a ritual act that negates the negation.

This may be the closest I can come to magic, by this retraction
to erase everything I can erase—to erase all I have written,
including this statement—and leave only whatever these words
have triggered or catalyzed in *your* brain (however untraceable
back through this text all of it would have been anyway).
For which it seems that *you,* dear reader, must take the credit.

You may think it a pose, but I assure you I only live for *this moment*—
the occult, forward-and-backwards-time-traveling, opalescent phantom
flurrying butterfly tornado in the process of coming into focus
between the two moments you and I both call *this moment.*

ASIDE: Obscurity

It is easy to see how Blake would appeal to authors with few readers—
a category to which I belong, but arguably one that, in the big picture,
also includes something pretty close to *all authors.*

Rather than compromise for a larger readership,
one may even aspire to belong to this category:
as Herman Melville put it, "it is my earnest desire
to write those sort of books which are said to 'fail.'"

Or I could say: hardly anybody reads my books,
but hardly anyone read Blake, and look at him now,
"famed in heaven," just as he always said he was!

Is Blake the patron saint of obscurity? And Butler, in a related
sense of the word? Is Woolf's worthwhile "work in obscurity"
a rationalization of marginality? Yes, so I maintain! *And?*

It's easy to think that poet John Keats was merely striking a pose
when he wrote of "the solitary indifference I feel for applause."

Is it believable that his pleasure in the moment of writing
would not have been compromised "even if my night's labours
should be burnt every morning, and no eye ever shine on them"?

After all, this was a man who had also written—prophetically,
as Blake claimed heavenly fame, and with almost as little reason—
"I think I shall be among the English Poets after my death."

It's harder to believe that he's posing when he strikes the metapose
of a poseur: "but even now I am perhaps not speaking from myself:
but from some character in whose soul I now live." Slippery fellow!

Keats followed through on his obscurity claims. He instructed,
from his deathbed, that his name not appear on his tombstone,
but only the words "here lies one whose name was writ on water."

His friend Joseph Severn thought these deathbed instructions came
from "the bitterness of his heart at the malicious power of his enemies."
Perhaps they do echo the defiant tone you can sometimes catch in Keats.

It reminds me of the joke about a New Yorker who, captured
by a headhunting tribe and told that a canoe will be made from his skin,
begins stabbing himself with a fork, shouting "Here's your fucking canoe!"

But if, in all of this, Keats was making a virtue of necessity
and collaborating in his own neglect or negation,
he was also *practicing the magic art of letting go*.

This is *what enabled him to remain in the moment of writing*
in order to make his poetry a kind of Ouija board for music otherwise
beyond human capability to compose (*negative capability,* he called it).

This is what enabled him to follow the music *into the future*,
"past the near meadows, over the still stream;/Up the hill-side"
to where it came to be "buried deep/In the next valley-glades."

ASIDE: Butterfly Tornado

And finally: one night, as I was finishing this book, I had a dream.
A boisterous, wolflike creature was dogging me, scaring everyone,
but I knew I could enforce some kind of social contract with him.

In the middle distance, a huge and snarling, lashing, toothy, furry
dog-headed cobra faced off against an opalescent human griffin
and an almost invisible scurrying, swirling thing, the wolf-dog's prey.

I directed him to chase it, hoping to relieve myself, in the process,
of his exhausting company, but he just shook his head and said aloud
he wouldn't dream of getting close to the scarier version of himself.

So we ambled off and came upon a house inhabited by William Blake,
partly boarded up, and peering in, I saw the mundane furnishings
of his life, and I saw some ghostly form of Blake, puttering about,
or possibly I just *pretended to see him*, to somehow fool the dog.

These possibilities were not opposed but overlapping, much as Blake
pretended (in *Marriage of Heaven & Hell*) he had dined with prophets,
but in the process of pretending somehow did commune with them.

As I dreamed, and again as I wrote the dream out, I had the sense
that something particular I couldn't make out was being whispered
in my ear—that I was being slipped a gift I couldn't quite open.

Searching, I noted subtle repetitions-with-a-difference:
the scary wolf-dog and the bigger, scarier cobra-dog;
the invisible, scurrying thing and the spirit of Blake,
also a kind of localized ripple in the visual field.

Then, later, the dream came back to me as a resolution of this book:
a tornado, pulling writer and reader and pieces of past, present, and
future into its orbit, forming a system in which the book participates.

Later, by accident, after a student on the first day of class
introduced herself as *Beatrice*, it came to me that the dream
elaborates a specific image from Blake's illustrations
to *Paradise Lost*: "Beatrice Addressing Dante from the Car."

In the *Purgatorio*, Dante, looking across a river to the earthly paradise,
witnesses a mystical pageant that includes Ezekiel's vision of a chariot,
where Beatrice appears to transport him up and onward on his journey.

The vision of the chariot, known as the *Merkabah*
(elaboration of the verb meaning *to ride; a vehicle*),
was among the holiest themes in Jewish mysticism;
all but adepts were warned against contemplating it.

The Merkabah informed Blake's vision of a fractally complex
universe of "wheel within wheel," which he set against
a machinic, wheel-outside-of-wheel universe of grinding gears.

Maybe the dream is a kind of endnote or afterthought, saying, in effect,
that all I've written might as well be commentary on Blake's illustration,
my bit of marginalia in a hermeneutical tradition of religious texts.

Or I could say, following Freud, that the dream and the vision were,
like all dreams, a coded and condensed way of embodying something
so revelatory it might otherwise blow the circuits of my consciousness.

Such wisdom might be represented by the two dangerous dogs,
the one that might be domesticated just enough to be integrated
into daily life, the other one might only safely glimpse from afar.

Or I could say instead: William Blake came to me in a dream.
He led me through scenes of the Merkabah before we returned
to his cottage where, tending the fire, he offered friendly advice:

It has been said that 'Whoever ponders on four things,
it were better for him if he had not come into the world:
what is above, what is below, what is before time,
and what will be hereafter,' but I say, 'If the Spectator
could Enter into these Images approaching them
on the Fiery Chariot of her Contemplative Thought—
if she could make a Friend & Companion of one of these Images
of wonder—then would she arise from her Grave & be happy.'

Acknowledgments

I am especially grateful to Sher Doruff, Manuela Rossini, and Masja Horn from Brill for supporting this book; to my co-teacher and magic-worker Melissa Buzzeo for encouraging me to persist in my folly; to intrepid copy editor Anne Canright; Chris Jensen for critical reading of Chapter 6; Andrew Barnes for his advocacy and example; Peter Patchen, for making the bank-shot diagram in Chapter 4 (and, he would want me to mention, making the actual bank shots); John Michael Greer, for pointing me to Vico; and Marty Babits, for standing with me in the spaces.

This book also comes out of a nexus of relationships for which I will always be grateful: to my co-teachers Youmna Chlala, Duncan Hamilton, Jennifer Miller, and Kristin Pape, for bringing me to the happy edge of my comfort zone; for all the students who have inspired me and made me feel that, with so much intelligence and humanity and humor, maybe the world has a chance (standing for all of them, I might just mention Holly Adams, Sonia Arora, Jonathan Ellis, Celina Hung, Jiang Wentao, Diana Joh, Rachel Ellis Neyra, Nina Foster, Naomi Frank, Gabrielle Gilbert, Ashley Kolodner, Xinan Ran); and to my dear friends and families: Alexandra Chasin, Jack Halberstam and Macarena Gomez-Barris, Dave and Mary Sandberg, Jim Adams and Nancy Whitley; Lang and Liam Walsh, Nikki Crook, Maggie and Sarah Livingston—and to Iona Man-Cheong—for teaching me how, as the old ballad says, "love is longer than the way"—and so in deep and sometimes wild optimism to Echo and Nova, Clio and Isabella, Zoa and Gray, Renato and Ixchel.

References

Chapter 1: Introduction

The-Two-Million-Dollar Comma: see Zachary Crockett, "The Most Expensive Typo in Legislative History," Oct. 9, 2014, http://priceonomics.com/the-most-expensive-typo-in-legislative-history.

When French sailors met the Yamacraw Indians: Peter H. Wood, "Circles in the Sand: Perspectives on the Southern Frontier at the Arrival of James Oglethorpe," in *Oglethorpe in Perspective: Georgia's Founder after Two Hundred Years*, ed. Spalding Phinizy and Jackson Harvey H. (Tuscaloosa: University of Alabama Press, 1989), 10.

Arthur C. Clarke's famous maxim: from "Hazards of Prophecy: The Failure of Imagination," in Clarke, *Profiles of the Future: An Enquiry into the Limits of the Possible* (1962, rev. 1973; London: Orion/Phoenix reprint, 2000), 36.

Aphorisms on Man/"*This should be written in gold letters*": Lavater Johann Kaspar, cited in Blake William, *Complete Poetry and Prose of William Blake*, ed. Erdman David, rev. ed. (New York: Anchor/Doubleday, 1988), 584. Subsequently cited as Blake, *Complete Poetry*.

Magic is the art of producing changes in consciousness: variously attributed to Aleister Crowley, Dion Fortune, and W.E. Butler.

Bruno Latour calls this *Science Two*: Bruno Latour, from the lecture series "Facing Gaia: Six Lectures on the Political Theology of Nature" (Gifford Lectures on Natural Religion, Edinburgh, Feb. 18–28, 2013), available at http://macaulay.cuny.edu/eportfolios/wakefield15/files/2015/01/LATOUR-GIFFORD-SIX-LECTURES_1.pdf, esp. 24–35.

André Green described emotion: in *Le discours vivant: La conception psychanalytique de l'affect* (Paris: Presses Universitaires de France, 1973).

In a 1972 lecture, climatologist Edward Lorenz: "Predictability: Does the Flap of a Butterfly's Wings in Brazil Set Off a Tornado in Texas," presented to the 139th meeting of the American Association for the Advancement of Science, available at http://eaps4.mit.edu/research/Lorenz/Butterfly_1972.pdf.

Erwin Schrödinger devised the famous thought experiment: first published in English in "The Present Situation in Quantum Mechanics: A Translation of Schrödinger's 'Cat Paradox' Paper," trans. Trimmer John, *Proceedings of the American Philosophical Society* 124 (1980): 323–38; available at www.tuhh.de/rzt/rzt/it/QM/cat.html.

Maxwell's Demon: James Clerk Maxwell first mentioned the demon in a letter of 1867 before publishing the scenario in his 1872 book *Theory of Heat* (repr. Mineola, NY: Dover, 2001).

No Atheists in Foxholes: Héctor Tobar, *Deep Down Dark: The Untold Stories of 33 Men Buried in a Chilean Mine, and the Miracle That Set Them Free* (New York: Farrar, Straus & Giroux, 2014).

"a complex system made up of many heterogeneous ... subsystems": Adam Sheya and Linda B. Smith, "Development through Sensorimotor Coordination," in *Enaction: Toward a New Paradigm for Cognitive Science*, ed. Stewart John, Gapenne Oliver, and Di Paolo Ezequiel A. (Cambridge, Mass.: MIT Press, 2010), 123.

REFERENCES

Blake on Firm Perswasion: from "Marriage of Heaven and Hell," in *Complete Poems*, 38–39.
Tom Paine called aristocratic titles "circles...": in *The Rights of Man*, joint edition with Edmund Burke's *Reflections on the Revolution in France* (New York: Doubleday/Anchor, 1973).
Virginia Woolf's famous book-length essay *A Room of One's Own*: quotes are from the Harcourt (New York, 1981) edition, with introduction by Mary Gordon; including the citation from Woolf's diary (vii).

Chapter 2: Complex Systems in a Nutshell

As Bob Dylan said, "You must leave now": Bob Dylan, "It's All Over Now, Baby Blue," from the album *Bringing It All Back Home* (Columbia Records, 1965).
Steve Shaviro: *The Universe of Things: On Speculative Realism* (Minneapolis: University of Minnesota Press, 2014).
Plant Sorcery: the basic facts are taken from Veronique Greenwood, "Eye of the Beholder," *New Scientist* 226, no. 3017 (April 18, 2015): 40–43; the plants-as-engineers account is my own.
"meshwork of selfless selves": Francisco Varela, "Organism: A Meshwork of Selfless Selves," in *Organism and the Origins of Self*, ed. Tauber I.A., Boston Studies in the Philosophy of Science, vol. 129 (Dordrecht: Springer, 1991), 79–107.
"Up Around the Bend": by John Fogerty, released as a Creedence Clearwater Revival single on Fantasy Records, 1970.
What Northrop Frye called the "Green World": in Frye, *The Anatomy of Criticism* (Princeton, N.J.: Princeton University Press, 1957), 182–84.
What Bakhtin called "Carnival": in Mikhail Bakhtin, *Rabelais and His World* (1941), trans. Iswolsky Helene (Bloomington: Indiana University Press, 2009), 303–436.
The precedence of *the aesthetic*: Shaviro, *The Universe of Things*.
"consumed with that which it was nourished by": from Shakespeare's Sonnet 73.
Sexual reproduction ... evolved along with programmed death: see William R. Clark, *Sex and the Origins of Death* (Oxford: Oxford University Press, 1996).
Luhmann's model: Cary Wolfe, *Critical Environments: Postmodern Theory and the Pragmatics of the "Outside"* (Minneapolis: University of Minnesota Press, 1998), 65–66.

Chapter 3: Magic by Example

"Surely some revelation is at hand": W.B. Yeats, "The Second Coming," in *Collected Poems of W.B. Yeats* (1937; New York: Collier,1966).

In Bruno Latour's account: see Bruno Latour, *We Have Never Been Modern*, trans. Porter Catherine (Cambridge, Mass.: Harvard University Press, 1993), 1–3, for a start.

"The dog starv'd at his master's gate": in Blake, *Complete Poems*, 490.

"We have never been modern": Latour, *We Have Never Been Modern*.

"the owl of Minerva takes its flight": G.W.F. Hegel, Preface to *Philosophy of Right*, trans. Dyde S.W. (Kitchener, Ont.: Batoche Books, 2001), 20; available online at http://socserv2.socsci.mcmaster.ca/econ/ugcm/3ll3/hegel/right.pdf.

In a Harvard Medical School study: reported in the *Guardian*, Dec. 22, 2010.

Your consciousness need only put the letter in the mailbox: Missy Vineyard, *How You Stand, How You Move, How You Live* (Boston: Da Capo Press, 2007), 158.

Meaning refers to the way one level of a system: for a more extensive account, see the chapter "What Is Meaning" from my 2015 book *Poetics as a Theory of Everything* (Poetics Lab Books; digital only).

Another—happier—Rube Goldberg sequence: this example is from my 2015 book *Poetics as a Theory of Everything*.

"A mother's laughter": *New Scientist*, July 17, 2010, 35.

The mirror box of neurologist V.S. Ramachandran: see V.S. Ramachandran and Sandra Blakeslee, *Phantoms in the Brain* (New York: Harper, 1998), 46–49, 52–55.

The most famous moment in the most famous role: the account of Garrick as Hamlet is drawn from Philip Ball's *Invisible: The Dangerous Allure of the Unseen* (London: Bodley Head, 2014); the quotations ("It made my flesh creep" and "no Writer in any Age...") are from p. 52. The analysis of the magic is mine.

Biting Game: D.W. Winnicot, *Playing and Reality* (1971; New York: Routledge, 1991); the quotations are from pp. 66–67. The analysis of the magic is mine.

Dog Whisperer: Sorry to say, I have not been able to locate the TV series in which this episode appeared.

Chapter 4: Future Perfect

Zen in the Art of Archery: by Eugen Herrigel, trans. Hull R.F.C. (New York: Vintage, 1999).

The novel *Caleb Williams*: William Godwin, *Things as They Are; or, The Adventures of Caleb Williams* (London: Penguin Classics, 2005).

Milan Rajković and Miloš Milovanović: "The Artists Who Forged Themselves: Detecting Creativity in Art," June 16, 2015, http://arxiv.org/abs/1506.04756; full text available at https://pdfs.semanticscholar.org/fbea/7887d563c04ff77e8a8df4265d8212ba263c.pdf.

"To Engrave after another Painter": letter to Dr. Trusler, Aug. 23, 1799, in Blake, *Complete Poetry*, 702.

Astronomer William Hartmann has a theory: the basic facts here are from Jacob Aron, "Christianity's Meteoric Rise," *New Scientist*, Apr. 25, 2015, 8–9; the interpretation is mine.

Chicago Times ... Times of London: cited in Ronald C. White, *The Eloquent President: A Portrait of Lincoln through His Words* (New York: Random House, 2007).

Regeneration through Violence: Richard Slotkin, *Regeneration through Violence: The Mythology of the American Frontier, 1600–1860* (Middletown, Conn.: Wesleyan University Press, 1973).
"As flowers turn toward the sun": Walter Benjamin, "Theses on the Philosophy of History," in *Illuminations* (New York: Schocken Books, 1968), 253–64.
"Mask of Anarchy": in *Shelley's Poetry and Prose*, ed. Reiman Donald H. and Powers Sharon B. (New York: Norton, 1977); the quotation is from p. 304.
Erica Chenoweth: see Chenoweth and Maria Stephan, *Why Civil Resistance Works: The Strategic Logic of Nonviolent Conflict* (New York: Columbia University Press, 2011).

Chapter 5: What is Religion?

The *Academia Secretorum Naturae* and *Accademia dei Lincei:* the account of these societies, and the quickie "history of cinema" account that follows, are condensed from Philip Ball's *Invisible: The Dangerous Allure of the Unseen* (London: Bodley Head, 2014).
Science also began to take shape *as a form of politics*: that's my soundbite version of Steven Shapin and Simon Schaffer, *Leviathan and the Air Pump: Hobbes, Boyle, and the Experimental Life* (Princeton, N.J.: Princeton University Press, 2011).
Renaissance mages "turned to number symbolism": Stanley Jeyaraja Tambiah, *Magic, Science, Religion, and the Scope of Rationality* (Cambridge: Cambridge University Press, 1990), 28–29. As the title of my own book indicates, I am indebted to and grateful for Tambiah's book—and would be pleased if I could claim a fraction of his elegance, insight, and erudition.
(as Eve Sedgwick put it), for some straight and gay people: Eve Kosofsky Sedgwick, *Epistemology of the Closet* (Berkeley: University of California Press, 1990), 25–26.
Halberstam suggests, the liberatory project is to *restore the chaos:* this is my paraphrase of Halberstam's point in his forthcoming book, *The Wild*.
Stephen Jay Gould proposed: in "Nonoverlapping Magisteria," *Natural History* 106 (March 1997): 16–22.
As Wittgenstein asked, if the purpose of the rain dance is to bring rain: cited in Tambiah, *Magic, Science, Religion*, 56.
Adrian Bejan has proposed what he calls *the constructal law:* Adrian Bejan and J. Peder Zane, *Design in Nature* (New York: Anchor Books, 2013).

Chapter 6: God 3.5B (A Nearsighted Evolutionary Panorama)

"The... focus": P.Z. Meyers, "Plant and Animal Development Compared," Feb. 17, 2008, http://scienceblogs.com/pharyngula/2008/02/17/plant-and-animal-development-c/.
"One cannot take too much care": Ludwig Wittgenstein, in *The Wittgenstein Reader*, ed. Kenny Anthony (Oxford: Blackwell, 1994), 286.

The Ancient Origins of Consciousness: by Todd Feinberg and Jon Mallatt (Cambridge, Mass. MIT Press, 2016); page numbers in subsequent citations refer to this edition.

Quasi-subjects/quasi-objects: see Bruno Latour, *We Have Never Been Modern*, trans. Porter Catherine (Cambridge, Mass.: Harvard University Press, 1993), esp. 51–55, 89, 139.

"Chuang Tzu said, 'See...'": Chuang Tzu, *Basic Writings,* trans. Watson Burton (New York: Columbia University Press, 1964), 110.

"Babylon throne gone down": Bob Marley, from the song "Rastaman Chant" (traditional, arranged by Marley, Peter Tosh, and Bunny Wailer [Neville O'Riley Livingston]) on *Burnin* (Island/Tuff Gong Records, 1973).

"climbing Mount Improbable": Richard Dawkins, *Climbing Mount Improbable* (New York: Norton, 1997).

Prehension: Colin McGinn, *Prehension: The Hand and the Emergence of Humanity* (Cambridge, Mass.: MIT Press, 2015).

André Green's definition of emotion: André Green, *Le discours vivant: La conception psychanalytique de l'affect* (Paris: Presses Universitaires de France, 1973).

As Lacan put it, "Meaning...": Jacques Lacan, *Feminine Sexuality: Jacques Lacan and the Ecole Freudienne*, ed. Mitchell Juliet and Rose Jacqueline (New York: Norton, 1982), 150.

"'It seems very pretty'": Lewis Carroll, *Through the Looking Glass and What Alice Found There* (New York: Macmillan, 1871), from Chapter 1, "Looking Glass House."

In Bruno Latour's terms, we would fall back: see Bruno Latour, "Facing Gaia: Six Lectures on the Political Theology of Nature" (Gifford Lectures on Natural Religion, Edinburgh, Feb. 18–28, 2013), available at https://macaulay.cuny.edu/eportfolios/wakefield15/files/2015/01/LATOUR-GIFFORD-SIX-LECTURES_1.pdf, esp. 25–51.

Freud linked "the antithetical meaning...": Sigmund Freud, "The Antithetical Meaning of Primal Words," in *The Standard Edition of the Complete Psychological Works*, trans. Strachey James, vol. 11 (1910; London: Hogarth Press, 1957), 155–61.

When groups prepare to face off: "Oxytocin Surge Preps Chimps for Battle," *New Scientist*, Jan. 14, 2017, 19.

"this is how we tried to love": Adrienne Rich, "Twenty-One Love Poems" XVII, in *The Dream of a Common Language: Poems 1974–1977* (New York: Norton, 1978), 34.

"the old signal sending noise at the wrong time": Wagner cited in Carl Zimmer, "Scientists Ponder an Evolutionary Mystery: The Female Orgasm," *New York Times*, Aug. 1, 2016.

"Contrast a termite castle": Daniel Dennett, "That's a Termite Colony between Your Ears," *New Scientist*, Feb. 11, 2017, 43.

The tattooist could tell: Stephanie Tamez.

I asked my doctor: Dr. Jeffrey Vieira.

How Wittgenstein describes philosophy: "The Nature of Philosophy," in *The Wittgenstein Reader*, 261–86.

The transculturally universal hallucination: see the account of hallucination in my *Poetics as a Theory of Everything* (Poetics Lab Books, 2015, iBooks edition), esp. 154–71.

"society of mind": Marvin Minsky, *Society of Mind* (New York: Simon and Schuster, 1988).

"many-headed" slime molds: see Steve Shaviro, *Discognition* (London: Repeater Books, 2016), Chapter 7.

Chapter 7: Blake Magic

"famed in heaven": from a letter from Blake to John Flaxman, Sept. 21, 1800, in Blake, *Complete Poetry*, 710.

Blake poems sung by Allen Ginsberg: many of these are available online at http://writing.upenn.edu/pennsound/x/Ginsberg-Blake.php.

The Fugs' rendition of "Ah, Sunflower" and "How Sweet I Roamed": from *The Fugs First Album* (Folkways Records, 1965).

The Doors of Perception: by Aldous Huxley (London: Chatto and Windus, 1964).

"if the doors of perception were cleansed": from *Marriage of Heaven and Hell*, in Blake, *Complete Poetry*, 39.

Smith's "my Blakean year": from the album *Trampin'* (Columbia Records, 2004).

She published a book of poems: Patti Smith, *Auguries of Innocence* (2005; New York: Ecco Press, 2008).

Alejandro Jodorowski's *Dune*: this account is taken from the documentary film *Jodorowski's Dune*, directed by Frank Pavich, 2014.

"Sales were less than brisk": from the online Blake Archive, www.blakearchive.org/exist/blake/archive/work.xq?workid=bb466&java=no.

"Auguries of Innocence" in Blake, *Complete Poetry*, 490–93; all subsequent citations of the poem refer to this source.

"scale is the result": Bruno Latour, "Facing Gaia: Six Lectures on the Political Theology of Nature" (Gifford Lectures on Natural Religion, Edinburgh, Feb. 18–28, 2013), available at https://macaulay.cuny.edu/eportfolios/wakefield15/files/2015/01/LATOUR-GIFFORD-SIX-LECTURES_1.pdf, 93.

"I wander thro' each charter'd street": in Blake, *Complete Poetry*, 26.

"The Man who can Read the Stars": in Blake, *Complete Poetry*, 769.

The "bourgeoisie produce their own gravediggers": Karl Marx and Friedrich Engels, *The Manifesto of the Communist Party*, in *The Marx-Engels Reader*, 2nd ed., ed. Tucker Robert C. (New York: Norton, 1978), 483.

"the arc of the moral universe is long": from King's 1967 speech "Why I Am Opposed to the War in Vietnam," available online (text and sound) at www.dailykos.com/story/2005/11/23/167357/--quot-Why-I-Oppose-the-War-in-Vietnam-quot-Dr-Maritn-Luther-King-Jr#.

"The Grey Monk": in Blake, *Complete Poetry*, 489–90.

"the master's tools": Audre Lorde, "The Master's Tools Will Never Dismantle the Master's House," in *Sister Outsider: Essays and Speeches* (Berkeley, Ca.: Crossing Press, 2007), 110–14.

"Fear & Hope are—Vision": in Blake, *Complete Poetry*, 266.

REM's "with love come strange currencies": from the song "Strange Currencies" on the album *Monster* (Warner Brothers, 1995).

The Beatles' "in the end, the love you take is equal to the love you make": Paul McCartney wrote the line, from the song "The End" on the album *Abbey Road* (EMI, 1969).

"One command, one joy, one desire": from *The Book of Urizen*, in Blake, *Complete Poetry*, 72.

As Tom Paine put it: in *The Rights of Man*, joint edition with Edmund Burke's *Reflections on the Revolution in France*, (New York: Doubleday/Anchor, 1973), 284.

Walter Benjamin's assertion: from "Theses on the Philosophy of History," in *Illuminations*, trans. Zohn Harry (New York: Harcourt, Brace, and World, 1968), 254–55.

"Neoliberalism's Stealth Revolution": Wendy Brown, *Undoing the Demos: Neoliberalism's Stealth Revolution* (Cambridge, Mass.: Zone Books, 2015).

"no other nexus between man and man": Marx and Engels, *Manifesto of the Communist Party*, 475.

"young girl's record of her own thoughts": Oscar Wilde, *The Importance of Being Earnest,* act 2.

"icy water of egotistical calculation": Marx and Engels, *Manifesto of the Communist Party*, 475.

"For the first time in history we've normalised walking speed": "Welcome to the Bionic Dawn," Hugh Herr interviewed by Catherine de Lange, *New Scientist*, Aug. 1, 2015, 24–25.

"To make that into a living formulation": Judith Butler, in George Yancy and Judith Butler, "What's Wrong with 'All Lives Matter'?" *New York Times*, Jan. 12, 2015.

Judith Butler, *Bodies That Matter: On the Discursive Limits of Sex* (London: Routledge, 1993).

Judith Butler, *Precarious Life: The Powers of Mourning and Violence* (New York: Verso, 2004).

"The Professor of Parody": Martha Nussbaum, in *New Republic*, Feb. 22, 1999.

"Listen to the fools reproach!": from *Marriage of Heaven and Hell*, in Blake, *Complete Poetry*, 37.

"They found him in the temple": Luke 2:46.

"Tlön, Uqbar, Orbis Tertius": story by Jorge Luis Borges, translated by James E. Irby, in *Labyrinths* (New York: New Directions, 1962), 18.

As Herman Melville put it: in a letter to Lemuel Shaw, October 6, 1849.

"the solitary indifference I feel for applause," "even if my night's labours," and "even now I am perhaps not speaking from myself": letter to Richard Woodhouse, Oct. 27, 1818, in *Letters of John Keats*, ed. Gittings Robert (Oxford: Oxford University Press, 1970), 157–58.

"I think I shall be among the English Poets after my death": letter to George and Georgiana Keats, Oct. 14/16/21/24/31, 1818, in *Letters of John Keats*, 161.

"the bitterness of his heart": Severn's words, from Keats's tombstone.

"past the near meadows": from "Ode to a Nightingale," in *Poems of John Keats*, ed. Stillinger Jack (Cambridge, Mass.: Harvard University Press, 1978), 372.

"It has been said that 'Whoever ponders on four things'": Mishnah, Hag. 2:1, cited in Gershom Scholem, *Kabbalah* (New York: Meridian/Penguin, 1978), 12.

"if the Spectator could Enter into these Images": slightly condensed (and, yes, I've altered the pronouns) from Blake's "Vision of the Last Judgment," in *Complete Poetry*, 560.

Index

aesthetic 23
affective labor 5–6, 20, 55, 117
Alexander Technique 53
Alice, in Wonderland 114–15
anarchy and anarchism 126, 130, 151
Ancient Origins of Consciousness 100, 111, 113–26, 130–37
Anthropocene 49, 89, 92, 106, 146
apology 158–59
attributive style 137
"Auguries of Innocence" 138, 141–46, 156
Austen, Jane 21–22
auto-ontological irreducibility 136
autopoiesis 26, 32, 34, 36, 40

bees 113
Bejan, Adrian 89
belief 18, 51, 53, 82
Benjamin, Walter 74, 150
bionics 152
Black Lives Matter 154–55
Blake, William 138–57, 161–63
   on Lavater 4
   on "firm perswasion" 17–18
   on engraving 68
   as a "four-ist" 94
Borges, J.L. 157–58
Bradbury, Ray 7–8
Brown, Wendy 151
Burghardt, Gordon 125
Butler, Judith 154–56
butterfly effect 6–9, 77

Caleb Williams 67
Chain of Being, Great 141–42
Chenoweth, Erica 77–8
cis-systems 105–6
clitoris 126–29
Clarke, Arthur C. 3, 4
cleavage 34
closure, operational 29, 32–3
coffee, revered by aliens as sacred 79
color, and color blindness 31–2
comma 2–3
complexity threshold 123–24

counterperformativity 63, 87
coupling, structural 29, 32–4, 36
criminalization 102

Dawkins, Richard 107
Dennett, Daniel 131–32
DNA 12, 110
dogs 61–62
dualism 94
Dune (film) 139–40

evolvability 109

Familia, Sagrada, La 131–32
Feinberg, Todd, *see Ancient Origins of Consciousness*
figurality and narrative 115
four-ism 94
Freud, Sigmund 120, 125, 146, 163

Garrick, David 56–8
Gaudi, Antoni 131–32
Gettysburg Address 63–4, 72–3
Godwin, William 67
Goldberg, Rube 3, 10, 43, 54, 57, 113
Gould, Stephen Jay 85
grammar 105, 118–19
"The Grey Monk" (Blake poem) 146–48, 153, 156
greyhound 21
Green, André 5, 112

Halberstam, Jack 85
Halliburton, David 93–4
hallucination 135–36
Hamlet 56–8
Hartmann, William 69–70
Heathrow 46–49
Herr, Hugh 152
hierarchy, tangled 37, 124–25
humankind, history of, in a single sentence 14

iii 33–34
intensity 114

interpositivity 27
intersectionality 1, 148
Irritable Bowel Syndrome 50–52

jabberwocky 114–15
Jodorowski, Alejandro 139–40
Journey to the West 125–26
Jupiter's Red Spot, and modernity 48

Keats, John 160–61
keystone species 144
King, Martin Luther, Jr. 145

Lacan, Jacques 114
Latour, Bruno
    Nature One and Two 5, 90, 116
    on modernity 49–50
    on scale 142
laughter 54
Lavater, J.K. 4
Law and Order (TV series) 101–3
learning 108–9
Lincoln, Abraham 63–4, 72–3
"London" (Blake poem) 144
Lorenz, Edward 6–9, 12
love 19, 137

Mallatt, Jon, *see Ancient Origins of Consciousness*
Marley, Bob 106
Marx, Karl 152
"Mask of Anarchy" 76–7
Maxwell's Demon 12–13, 35
meaning 15, 53–4, 98–9, 114, 120, 129, 141
merkabah 163
melancholy, cosmic 80
metacognition 22
meteors 69–70, 110
Meyers, P.Z. 97
Milovanović, Milos 68–9
Minerva, Owl of 50, 91
Minsky, Marvin 137
monism 94
monotheism 87, 150
mutation 40, 44, 126

narrative and figurality 115
naturalism 68

New Science (Vico) 91–6
nonlinearity, in writing 66–68
nonviolence 74, 77–8

Object-Oriented Ontology 42
Oedipus 145–46
olfaction, *see* smell
open systems 32
orgasm 126–29
othering 87
overestimation 34–5

pain, chronic 132–34
Paine, Thomas 18, 150
panpsychism 43
pattern 41
Paul, Saint 69–70
performativity 70, 88, 143, 159
personification 39
phallic motherhood 148
phantom limbs 54–6, 58–9
phantom vibration syndrome 56
Phenomenological I 104, 114
photosynthesis 30
Pierce, C.S. 93
pool 12, 64–6
pregnancy 52–3
prehension 107
primal words 120
providence 95–96

rain dancing 87–8
Rajković, Milan 68–9
Ramachandran, V.S. 54–5
referentiality 88, 124, 130, 159
reflex 124
Regeneration Through Violence 73
relativism 86–7
relaxation 62
religion, as "binding back" 11
rhizome 84
A Room of One's Own 19–22

Schrödinger, Erwin 9–12
Sedgwick, Eve 85
selection 109, 114, 128
self-consciousness 111
self-organization 68–9

self-referentiality 121, 124, 135–6
semiotics 107
sexuality 85, 127–28
Shaviro, Steven 25, 39, 137
Shelley, Percy 76–77
smell 107, 111–14
smoking, quitting 71
someness 16, 45, 119, 123
Smith, Patti 139
Speculative Historical Acupressure 74–5
Sprinker, Michael 139
sunflowers 42, 74, 112, 138
symbol 143
systemhood, primordial 101
"A Sound of Thunder" 7–8

tattooing 132–33
Terminator 2 (film) 103
thermodynamics, Second Law of 13
three-ism 94
time travel 75–6
"Tlön, Uqbar, Orbis Tertius" 157–58
tornados 6–8, 11, 33
transduction 99

trans-systems 105–6, 119
Triborough Bridge 105
2001: A Space Odyssey (film) 103, 104

underestimation 34–5
unity 37, 124–26, 130
The Universe of Things 25, 39

Varela, Francisco 37
Vico, Giambattista 91–6

Wagner, Gunther 126
Whig history 95
Wilde, Oscar 152
wildness 23, 44, 142
will 17–19, 21
Winnicot, D.W. 59–61
withness 45
Wittgenstein, Ludwig 88, 97, 134
Wolfe, Cary 43
Woolf, Virginia 19–22

Yiddishisms 12, 34

www.ingramcontent.com/pod-product-compliance
Lightning Source LLC
Chambersburg PA
CBHW060304010526
44108CB00042B/2660